MW01633707

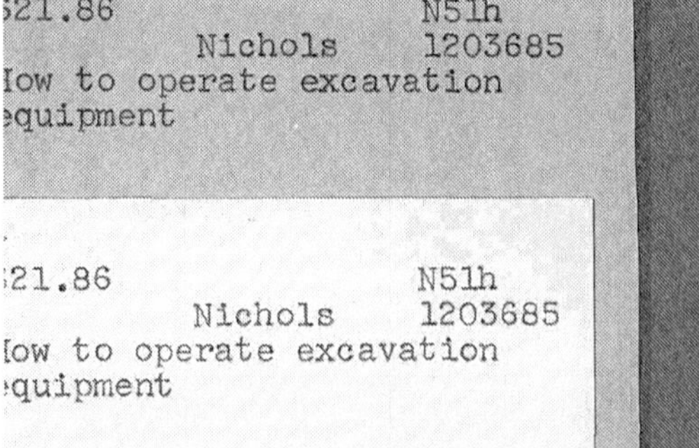

621.86 N51h
Nichols 1203685
How to operate excavation equipment

HOW TO OPERATE EXCAVATION EQUIPMENT

Written

to help the men and the machines that move the earth

and to instruct and entertain those

who like to watch them do it

HOW TO OPERATE EXCAVATION EQUIPMENT

BY

Herbert L. Nichols, Jr.

ILLUSTRATIONS BY HELEN SCHWAGERMAN

NORTH CASTLE BOOKS

GREENWICH · CONNECTICUT · 1954

This book is now published by
D. VAN NOSTRAND COMPANY, INC.
120 ALEXANDER STREET
PRINCETON, N. J.

FOREWORD

We feel it is a great honor to have been allowed to read the advance proofs of "How to Operate Excavation Equipment."

The construction industry, and particularly that part of it concerned with excavation, cannot help but be grateful to the author for putting into writing the basic facts about operation of the machinery which makes its accomplishments possible. Now that this writing has been done, it seems hardly credible that such important information has, up to now, been conveyed almost entirely by word of mouth, or has been learned the hard way when instruction was lacking. And often this way has been very hard.

The author is well qualified by his background as both an operator and a contractor to speak with authority, and his material is explained clearly and simply. It seems likely that its readership will go beyond the operators and contractors for whom it was written, to the large public for whom we put windows in our barricades so that they can watch machines at work.

For these people, and for their sons from among whom the next generation of construction men will come, this book opens a wider window, so that they can see not only what we do, but how we do it.

F. H. McGraw & Company
George J. O'Connell
Equipment Manager

(F. H. McGraw & Co. is the prime contractor on the billion dollar atomic project in Kentucky.)—ED.

PREFACE

"How to Operate Excavation Equipment" is a serious book of instruction, written to help the operator and contractor. It is authentic and reliable, the product of nineteen years of work with machinery. It has been checked in every detail by contractors, operators, and equipment manufacturers.

Yet it is also the absorbing answer to the healthy curiosity of millions of men and boys—a how-to-do-it book to make all its readers more aware of the fascination of earth moving, and to start many of them in worthwhile and well paid careers.

It supplies detailed non-technical information about how to run the important types of excavating, hauling, and grading machinery. Details of mechanical design and of specific jobs will be discussed in later books.

Operation techniques vary somewhat among different operators, and in different parts of the country. The methods chosen for explanation are the basic ones which are easiest to learn, least likely to damage the machine, or are best known to the author. After an operator has become thoroughly familiar with one way to run a machine, he can readily adapt himself to other methods.

The important subject of lubrication is not discussed because of differences in requirement, and the fact that a lubrication schedule is available for almost any machine. But the novice operator should bear in mind that neglect of lubrication can do serious and expensive damage very quickly, and will even put a good machine on the scrap pile.

No attempt is made to list the hundreds of differences in arrangement of controls among different makes and models. But once a man is properly introduced to a machine, this book will tell him what he needs to know.

But remember—these earth movers are not toys. They are heavy, powerful, and expensive. If mishandled they are dangerous to the operator and to everyone around him, and can be fearfully destructive to property and to themselves. The operator should respect them always, be slow and careful in his actions, and should never allow his confidence to exceed his skill.

The author is sincerely grateful to the equipment manufacturers, contractors, engineers, foremen, and operators who have supplied information and checked manuscript for this book.

H.N.

CONTENTS

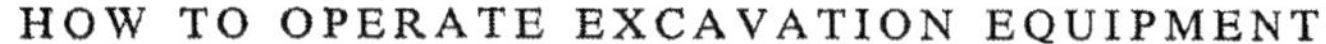

HOW TO OPERATE EXCAVATION EQUIPMENT

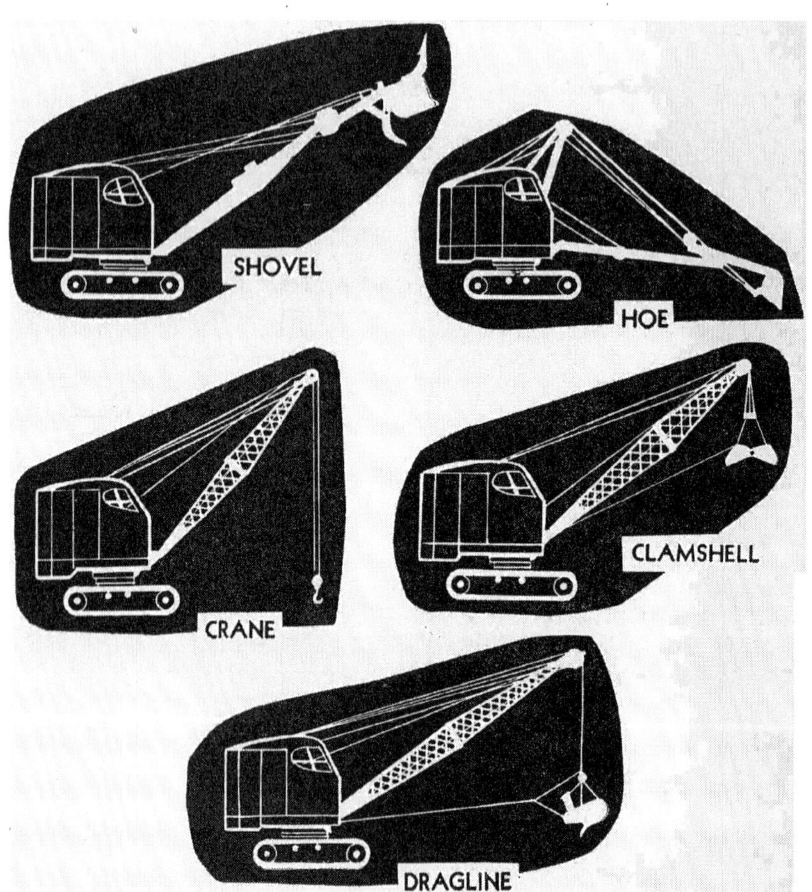

Fig. 1. Five shovel rigs

CHAPTER ONE

REVOLVING SHOVELS

Revolving shovels made their first appearance in 1835 in the form of a part swing dipper shovel mounted on railroad tracks. It was powered by steam, it was slow and clumsy, but it did the work. As the years went on, shovels gradually became stronger, faster, and lighter. During the last fifty years they have sprouted a variety of new attachments and have left the rails to walk on crawler tracks and rubber tires. Every year they encounter more competition from other kinds of machines, but they are still the basic tools for earth moving.

Each shovel is made in three structural divisions. Anatomically, the top or revolving unit is the head and torso, the mounting or travel unit is the legs, and the various attachments are the arms and hands. A revolving and a travel unit together make up a "basic shovel."

There are five attachments (also called rigs or fronts) which have primary importance. These are known best under the names dipper shovel (or dipper stick), backhoe, dragline, clamshell, and crane. They are shown in Figure 1.

THE BASIC SHOVEL. Power Trains. A basic shovel includes three sets of machinery. One of them is made up of deck-mounted drums which are fitted with spools for cable or with sprockets for chain, and which are rotated and stopped by clutches and brakes. These control the in-and-out and up-and-down movement of the bucket for digging and dumping.

The second set rotates (swings) the deck, upper machinery, and front end around a hollow center pin. The upper unit is supported by rollers running in a circular track or turntable, and is rotated by a vertical shaft and a gear which meshes with bullgear teeth in the turntable, and "walks" around on them, swinging the shovel as it does so. The gear is controlled by a reversing clutch (a unit consisting of two friction clutches and a set of bevel gears), which can turn it in either direction.

This mechanism is an important factor in the efficiency and adaptability of the shovel. It enables it to face in any direction for digging and dumping, and to move loads quickly anywhere within its reach. Comparatively few wearing parts are involved, none of them work in the dirt, and friction losses are slight.

The third power train provides means to walk or propel the shovel. A vertical shaft extends downward through the deck and the hollow center pin to drive a horizontal axle, which has a clutch and a brake on each side. These may be of either jaw (tooth) or friction construction. A pair of sprockets on the ends of the outer axles drive the tracks through roller chains.

The propel mechanism is controlled by a reversing friction clutch. This is usually the same unit that controls the swing. It can be connected to either the swing or the propel mechanism by shifting of jaw clutches. In some machines propelling is done by the crowd-retract clutches of the digging machinery, also by means of a jaw clutch shift. In truck mounted shovels the propel mechanism is usually absent or disconnected.

Small shovels use gasoline, diesel or electric engines, with mechanical drive to all moving parts. Large shovels may use diesels, with mechanical or electric drive; or several electric motors supplied from "high lines" through a cable.

The engine is fitted with a disc clutch, so that it can be cut off from the machinery. It may also have a fluid clutch to cushion it against shocks and to prevent stalling. A torque converter can be used in specially designed machines, for the same purposes and to give added power in heavy pulls.

Shovels of three quarter yard size and smaller may have an engine transmission with two or more speeds. The high speed is intended for traveling, or exceptionally light and open work. In addition, the governor may be provided with a special setting to speed the engine while walking.

Controls. The operator's seat may be on either side of the deck, forward of the center. Levers and pedals controlling all functions of the machine are near it. A representative set of controls is shown in Figure 2.

In this machine, the hoist lever engages the hoist clutch when pushed forward. The lever is held back by a light spring, but when pushed forward beyond the point necessary to engage the clutch, it locks over center, and requires a pull to disengage it. Use of this lever rotates the hoist or front drum so as to wind in the cable attached to it.

The digging, drag, or crowd lever when pushed forward engages the clutch of the same name, winding in cable that forces the bucket into material being dug. If used with a dipper stick it can be pulled backward from center also, engaging the retract clutch. These movements are made against spring tension, and both clutches lock over center.

The boom hoist lever, when pulled back,

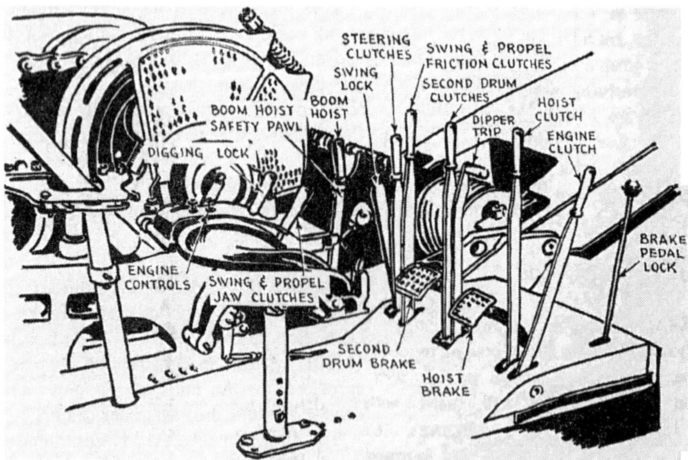

Fig. 2. Shovel control station

engages the boom clutch, causing the boom cable to wind in and raise the boom. In center position, it releases this clutch and allows an automatic brake to lock the boom drum. When pushed forward, it releases the brake and allows the weight on the cable to turn the boom drum as fast as the drum shaft turns. There is also a ratchet which can be set to prevent the boom cable from paying out at all.

The swing lever, when pushed forward, engages the front swing clutch and causes the shovel to swing left or, in walking position, causes it to move forward. When pulled back, this lever engages the other swing clutch, with reversed results.

The swing lock, when pulled back, engages a toothed lock with the bull gear and prevents the shovel from swinging. The swing-propel shift lever, when forward, engages the jaw clutch for swinging. When back, it connects the walking jaw. In the middle both are disengaged.

The hoist brake locks or slows the rotation of the hoist drum, the drag brake does the same for the drag drum. These brakes are held in released position by springs, but can be locked down by engaging with a ratchet. Pushing on the top of the pedal locks it, on the bottom releases it.

The throttle regulates the speed of the engine by changing the governor setting. The steering lever is used only when walking. Forward turns to the right, backward to the left, center straight ahead, in forward motion.

The engine clutch is used to disconnect the machinery from the engine, for safety when adjusting or lubricating with the engine running, and for starting in cold weather.

The digging lock will lock the shovel tracks when pushed forward, will release them when pulled back, and in the center will allow the shovel to more forward only.

Controls are different in each make of shovel. Many use vacuum, air, or hydraulic controls, with shorter operating levers. Steering brakes are generally used to hold for digging also, and may be applied by hand or automatically. The direction of throw of operating levers can often be reversed.

Lubrication. Most of the deck machinery may be enclosed in oil tight cases, and lubricated by gears dipping oil out of one or more reservoirs and carrying it to distribution points, from which it flows over the other gears and bearings and back to the reservoirs. The oil level should be checked frequently by removing inspection plugs, and sediment should be drained from the bottom sumps occasionally.

The turntable gear is lubricated by pouring heated crater compound through a hole in the floor over the gear, while the shovel revolves. This gear should be inspected at least once a day, and greased when bright spots appear on the teeth.

Pressure gun fittings throughout the shovel are greased in accordance with a schedule supplied with the machine. In general, bushings require more frequent attention than anti-friction bearings, some of which may be sealed so as not to require lubrication. The crowd chains for the dipper stick should be kept painted with oil or light grease, but the tracks and drive chains should be left dry.

If the deck machinery is not in enclosed cases, the gears are usually greased with crater compound. Small applications are recommended, as surplus works down to the floor, where it may combine with dirt to build up hard deposits which will obstruct the horizontal gears. If it gets in the underbody, it may cause jaw clutches to stick, and spoil traction brake linings.

Steering. If both axle jaw (steering) clutches are engaged, rotation of the vertical propel shaft will turn both tracks equally. If the right clutch is disengaged, the left track only will drive. The effect on the shovel's direction of travel will be determined by the steepness or roughness of the footing, and friction in the right track. See Figure 37. On firm level ground the shovel will move straight forward, or a little to the right, on steep upgrades or on rough ground, it may turn quite sharply to the right. If the right axle is locked by the brake, the right track will stop, and the left one will walk in a circle around it, turning the shovel very abruptly. Turning in soft material is difficult, as the locked track digs in. On slippery surfaces one or both tracks may skid, preventing accurate steering.

Turning to the left involves engaging the right hand clutch and disengaging the left one, and perhaps locking the left side if a sharp turn is desired. A gradual turn may be made by use of the clutch alone, by partial application of a friction brake, or by a succession of short, sharp braked turns with straight runs between.

If the machine is moving down a grade steep enough so that its engine is holding it back rather than pushing it forward, the steering action of the clutches will be reversed; that is, disengaging the right clutch will cause the shovel to turn to the left, as the disconnected track can roll faster than the one connected to and held back by the engine. When the brake is used, steering is about normal regardless of grade.

The shovel should be stopped before engaging or disengaging the jaw clutches or brakes. While they are transmitting the engine power to the tracks, they are almost impossible to disengage; and when apart they might be damaged by being forced into engagement while the jaws are turning at different speeds. The shovel is stopped by partially engaging the friction clutch that would move it in the opposite direction or by using friction brakes. In an emergency a ratchet lock or jaw brakes can be used for stopping but this is very jarring to the machinery.

The steering method is the same going backward (toward the bull wheels), except that for a turn in the same compass direction the opposite clutch and brake are used.

Swing Lock. It is usually necessary to lock the shovel from swinging before it can be walked safely. This is partly because the power is applied to the vertical propel shaft from above the turntable, and carried to the live axle below it. This power should turn the shaft and axle, but if the resistance of the axle is sufficient, the shaft will remain stationary and the revolving unit rotate around it. In easy walking, the swinging tendency will be slight, in soft or uphill travel the power will all go into swinging unless checked.

The swing is usually locked by means of a toothed bar anchored to the bottom of the deck, which can be engaged with the turntable gear by means of the swing lock control lever. A friction swing brake is used in some machines.

If the shovel is equipped with separate reversing clutches for propel and swing, the deck and boom can be kept in line or swung to avoid obstructions while walking by use of the swing clutches. However, it is usual for the operator to lock the swing for extended moves in the open, so that he will not have to keep alert to vary the engagement of the swing clutch if the propelling torque changes.

Steering Controls. The steering clutches and brakes may be operated by a single lever, which engages both clutches when in middle position, moves forward to disengage the right clutch, and backward to disengage the left, and engages the respective brakes by further movement. The control linkages for these, and for traction ratchets, are in the hollow center of the vertical propel shaft, as that is the only place where they would not be sheared off as the shovel revolves.

If the steering brakes are used to hold the shovel while digging, separate levers are used for right and left hand steering brake and clutch. The brake is applied and the clutch released at one end of the lever travel, and the brake released and the clutch engaged at the other end.

Tracks. Reference is made to Figure 3. If the shovel moves forward, the bull wheel will turn so that the teeth will mesh with the pads under and ahead of it, picking them up one by one and passing them around its rear and then forward over its top. As the bull wheel moves forward on the track, it pushes the truck frame and the idler ahead. The idler, pressing against the track in front of it, pulls on both top and bottom sections, but since the bottom is pinned down beneath the shovel, and is pulled tight by the bull wheel, the idler pulls in the track from above about as fast as the bull wheel supplies it from behind, and turns it back to the ground, where it stays, supporting the truck wheels and weight of the shovel until picked up by the bull wheel again.

If the shovel is walking backward, the mechanism may behave in two ways. If the footing is good, and the grade nearly level, the bull wheel will pull in some slack from the top of the track and pass it underneath, moving backward and pulling the truck frame and idler after it. The track it has turned underneath supports the shovel and is then carried up by friction with the idler.

If the footing or grade is such that there is insufficient traction in the pads directly underneath the bull wheel to move the shovel, they will be skidded forward against the rear truck roller, the bull wheel will pull the upper track tight, and it will pull the idler backward. The idler moves truck frame, shovel, and bull wheel backward.

There are several reasons why a shovel or other tracked vehicle should be walked forward when possible. In forward motion, the track is laid smoothly along the ground, and its slack hangs harmlessly in gradual curves in the top section. The heavy strain is on the short piece from the bottom of the

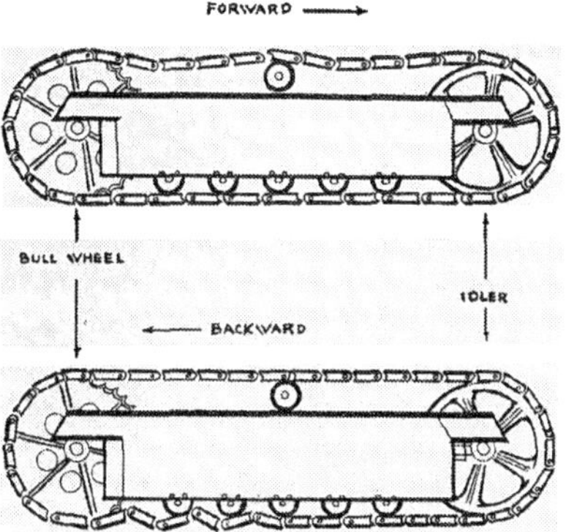

Fig. 3. Track behavior

bull wheel to the first truck wheel supporting a good share of the shovel's weight. In reverse movement, the track may be kinked in one or more damaging angles which the truck rollers must climb over or push down, and the traction stress is on the whole distance from the top of the bull wheel to the front slope of the idler, causing excessive wear and danger of breakage.

Track pads are available in several widths. Wide tracks hold the machine up better on soft ground, but they somewhat increase the effort of steering, and are more subject to severe twisting stresses on uneven ground. They are heavier and more expensive, and cause complications by increasing the over-all width of the machine.

Extra long tracks increase stability, and are desirable for long boomed machines, and those which must handle heavy loads.

Flotation can be greatly increased by placing supporting platforms on the ground.

Both the tracks and the drive chains wear, and need periodic adjustment. Bull wheels and idlers are mounted so they can slide backward and forward on the truck frame fastenings, and the position of each is fixed by heavy bolts. Lengthening the front bolts will force the idler forward and tighten the track only, lengthening the rear bolts will force the bull wheel and sprocket backward and tighten both the track and the drive chain. Care must be taken not to turn a wheel sideward by unequal adjustment of paired bolts, as it will then tend to walk out of the track.

If the track is allowed to become loose, it may come off when working or making turns on uneven ground. Slack also allows the shovel to rock back and forth under digging pulls, wearing the track parts, and makes proper control difficult when climbing ramps onto trailers. A too tight track may be badly strained or break on rocks, and wear rapidly because of tension on the hinge pins.

Maximum crawler speeds range from about three quarters of a mile an hour for medium large machines, to four or more for small ones.

Rubber Mountings. The shovel revolving unit may be carried on a turntable fastened to a truck chassis. The truck engine is then used for traveling, and the shovel engine for digging. Ordinarily, the shovel controls for walking, steering, and traction braking are disconnected or missing. Occasionally, however, arrangements may be made to connect the shovel engine to the wheels through the truck transfer case, and to steer and brake the truck from the shovel cab. This mechanism is used for short moves at low speeds in the pit.

Any standard truck chassis of sufficient rated capacity can be used to carry a shovel, although considerable extra bracing is required, and better service should be obtained from a chassis specially engineered for the shovel. Tandem drive gives best support, and all wheel drive is advisable in work where mud or sand may be encountered.

The truck should have the right hand side of the cab cut away, to make it possible to carry the boom forward at a low angle. The frame and the rear axle should be a single rigid unit while digging. Springs may be eliminated, locked during work, or replaced by stabilizer bars in the tandem construction.

The truck mounted shovel can ordinarily swing in a full circle, but with most attachments can work through only 270° because of interference of the cab and the truck front.

Shovels may also be mounted on rubber-tired chassis of the same general dimensions as the crawler units, with two or four wheel drive from the shovel engine. Top speed is somewhat lower than the truck mounting, but maneuverability is better, and full-circle work is possible with most models.

Outriggers. Outriggers may be used to increase stability. These may be beams which can be slid or folded out of the bumpers, usually only at the rear, and which are supported by blocks or jacks. They provide a much larger and more rigid base than the tires. When they are used, lifting capacity is greater than that of a crawler of the same size, particularly when working off the back so that the truck engine acts as a counterweight.

Platforms may be put under the blocks to give support when working on soft ground, and the jacks may be used to lift the wheels out of mud holes.

Uses of Truck Mounting. The advantage of truck mounting is its capacity for rapid and inexpensive movement from one job to another. With many models, the shovel can be placed and locked in traveling position in less than a minute, and then moved along roads at twenty to thirty-five miles an hour.

On the job, however, the truck-shovel suffers from lack of maneuverability. Instead of turning in nearly its own length, as the crawler, it must have considerable space in which to turn or side-step.

Its most important weakness for excavation work is the ease with which it can get stuck. Even with all wheel drive, its ability to get in and out of soft spots is greatly inferior to that of a crawler, and with rear drive only, constant care must be exercised to keep it out of trouble in soft ground and during rains. This disability has little effect on its usefulness when equipped as a crane, since most of its work can then be done on pavement.

Walking Draglines. In this book, mention is frequently made of crawler machines walking, but this is with apologies to the true walking draglines (walkers).

These have a working base which consists of a wide, flattened cylinder called a tub, which rests on the ground and carries the turntable on its upper side. The walking mechanism is operated by cams or cranks, one on each end of a large diameter multiple-part shaft mounted across and extending beyond the sides of the upper frame and the tub. They are connected to the walking frames and the walking shoes. When the shaft is rotated the shoes contact the ground, raise the machine, and support its weight as a movement or step is made. The rear lifts clear, and the front drags.

The process is very similar to that by which a man on his hands and knees progresses by resting his weight on his hands and swinging his body forward with feet dragging, then rests on his knees and toes while he moves his hands forward again.

Steering is done by lifting the shoes and turning the revolving unit until its rear faces the desired direction, then stepping off as described. This is probably the only steering device in heavy equipment which takes negligible power, does not tear up the ground and permits turns of any desired sharpness under any footing conditions.

This easy steering and the light ground pressure made possible by the large area of the base are the special advantages of this mounting. They enable it to work in places dangerous or impossible to crawler machines because of difficult terrain, soft footing, or nearness to loose down slopes.

The disadvantages are that they are slow, with top speed around one fifth of a mile an hour, and are very wide, thirty feet or more. They must be disassembled to ship from one job to another, and are at a disadvantage in the bottom of a pit, both because of their width, and the fact that they must walk away from their boom, which might be blocked by a hill.

DIPPER SHOVELS

A half yard dipper shovel, with a diagram of its digging parts, is shown in Figure 4. The dipper shovel is the ideal machine for loading trucks. It is accurate because of the three-way control of the bucket and the ability to dump in any position. It is fast because only a small part of its weight—the stick and bucket—are involved in the digging motion, and because of its ability to dig in any spot it can reach with minimum waste motion approaching and leaving the cut.

The dipper is also the attachment for hard, heavy digging—hardpan, boulders, ledge, and blasted rock. The weight of the boom, backed to a varying extent by the weight of the shovel itself, holds the bucket to its work. The effective cutting angle of the teeth in a bank, aided by the variable direction of pressure provided by the hoist and crowd, enable it to cut highly resistant material, and to break up cracked and fissured rock formations. The shape of the bucket enables it to pick up objects much larger than itself without chaining. Larger size shovels are increasingly effective at this type of work.

However, it is necessary that the shovel be able to move into its work constantly in order to retain its effectiveness. Solid rock projecting from the pit floor will prevent getting at the bank unless ramps are built over the rock, or bulldozers used to push the bank to the shovel. In digging below the tracks, the downslope should not be much over 20 per cent if the machine is to follow the work. If it is

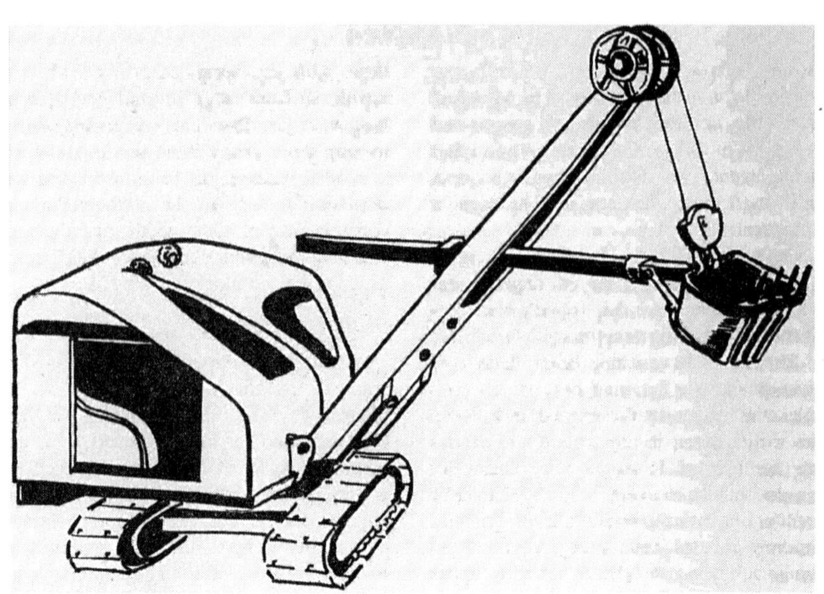

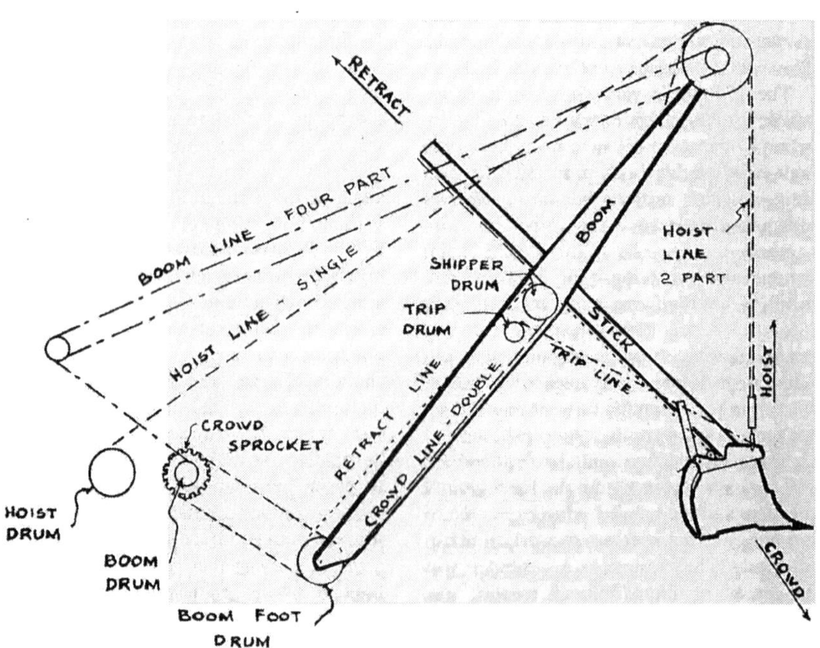

Fig. 4. Dipper shovel

to stand at the edge and dig, it can cut down steeply five to seven feet, but is limited as to width of cut, and unless digging against a bank may push away about as much material as it picks up. Narrow ditches can be dug effectively only in ground free from large boulders which would tear up the ditch sides, and firm enough so that the shovel can be worked straddling its ditch, with or without platforms.

Digging. Figure 2 shows the operator's station in a typical half yard shovel, and Figure 5 a sequence of movements in digging a bank.

The shovel is swung to face the section to be dug, with the hoist and crowd clutches disengaged and the brakes set. The stick is about horizontal with its center forward of the shipper drum. The hoist brake is released, the bucket falls in an arc, and the back of the stick rises, all motion being centered on the shipper shaft. The momentum of the bucket carries it back of the line of the shipper shaft and application of the hoist brake stops it before it hits the underside of the boom.

The controls cannot hold the bucket in this position and it will tend to swing down and forward by gravity. But as the hoist brake is applied, the crowd brake is released and the stick slides downward through the saddle block until the bucket teeth rest on the ground, as in (C).

It is now desired to move the bucket forward horizontally, to level irregularities on the ground, then to continue the level grade into the bank. This is done by releasing the hoist brake and engaging the hoist clutch, which pulls the bucket forward and downward in an arc around the shipper shaft. The crowd brake is released, and the retract clutch partially engaged so that it will pull the bucket up to compensate for the downward path of its arc, until the stick is vertical.

From here forward the hoist tends to pull the bucket upward. The bucket can now be kept to its level grade by releasing the retract clutch and partially engaging the crowd brake, so that the stick can slide down through the saddle block just enough to compensate for its upward curve from the hoist line. If the digging is hard, the crowd brake may be released and the crowd clutch engaged sufficiently to force the bucket against the dirt, but not enough to cause the boom to be raised by the crowding force.

As the teeth penetrate the bank, (E), the bucket fills. When it is full the crowd brake and clutch are disengaged, and the retract clutch is engaged so that the bucket is pulled back out of the bank while still hoisting, and the shovel is now ready to swing.

To summarize, the bucket was swung back into digging position by first releasing the hoist brake, then the crowd brake. For the digging, the hoist clutch was continuously engaged, while the depth and angle of digging were controlled by the crowd-retract mechanism, which would also allow digging up or down slopes or even vertical steps.

If the bucket threatened to strike the bank or the ground while being swung back, the retract could be used to lift it without interrupting its backward motion. It is generally not necessary to start the digging back at the tracks, and the bucket can be dropped at any point by releasing the crowd brake and retract clutch. The usual procedure is to drop it at the toe of the bank.

Dumping. This shovel swings to the left when the swing lever is pushed forward. Well before the end of the swing, this lever should be pulled into neutral, the swing continuing from momentum until stopped by pulling the lever back to engage the other swing clutch just enough to act as a brake. If the swing is to the right the lever is moved in reverse direction.

During the swing the position of the

bucket is adjusted by means of the hoist and crowd-retract controls so that it will be in correct position to dump as soon as it is over the truck.

If the truck is facing directly away from the shovel, the placement of the load in the front or rear of the body is regulated by the crowd-retract, placement from side to side by the swing, and the height above the body by the hoist; except that crowding beyond the boom point also raises the bucket if the hoist brake is held.

When the bucket is in correct position, the door in its bottom is unlatched by means of the dipper trip and swings down and back by gravity, allowing the contents to fall. If dirt sticks it can usually be dislodged by jerking with the crowd-retract lever, alternately engaging and disengaging both clutches. The load may be dumped in a heap or spread by moving the bucket as it dumps. The door is slammed shut by a quick downward or backward motion of the bucket. With the door closed the bucket can be used to tidy up the load on the truck, pushing and pulling the humps into the hollows.

Walking. As soon as the shovel has dug away so much of the bank that it does not reach enough to easily fill the bucket, the operator locks down the operating (hoist and crowd) brakes, engages the swing lock, disengages the swing jaw clutch and engages the walking jaw clutch by moving the swing-walk lever, then engages the forward swing friction clutch by pushing the lever forward, so that the machine moves forward. In his new position he releases the swing lever, disengages the walking jaw clutch, engages the swing jaw clutch, releases the swing lock and the operating brake locks, and is ready to work.

The shovel may be held in position against digging push by the ratchet lock on the live axle, which can be set to permit forward movement but hold against rolling back. If there is no ratchet it is necessary to set the traction brakes to dig and to release them to move. This might be done either manually or automatically, depending on the make and model of the machine.

Leveling. If the space into which the shovel is advancing is too rough to afford good footing it may be leveled by cutting off the humps with forward motion of the bucket, or by dragging the bucket backwards. It can be allowed to rest on the ground and be pulled back with the retract, releasing the hoist brake enough so that the bottom just brushes the loose material as it swings back.

A pit may also be leveled behind the shovel for the benefit of the trucks by holding the bucket in light contact with the ground and swinging the shovel. This will cause dirt to be pushed across the bucket marks smoothing them over, will pull wastage around the sides of a windrow back into digging reach, and will move loose rocks out of the way. It is a practice which is universally condemned by shovel manufacturers, as it imposes severe twisting strains on the stick, saddle block, and boom pins, shortening their life, or sometimes causing complete collapse. It will be seen that if the boom and stick are swinging with the momentum of their own weight and that of the shovel, and the bucket catches on an obstacle, the twist is severe.

However, no amount of disapproval will stop operators from using this technique as it is often the easiest way to finish work, and sometimes the only way. The twist may always do some damage, but if the front end is strongly built and the operator skillful enough to avoid contact with too-resistant material, the advantages will far outweigh the extra wear. No operator should do this, however, until he has acquired very precise control over the particular machine.

The shovel can also be swung while digging in order to square off a bank or

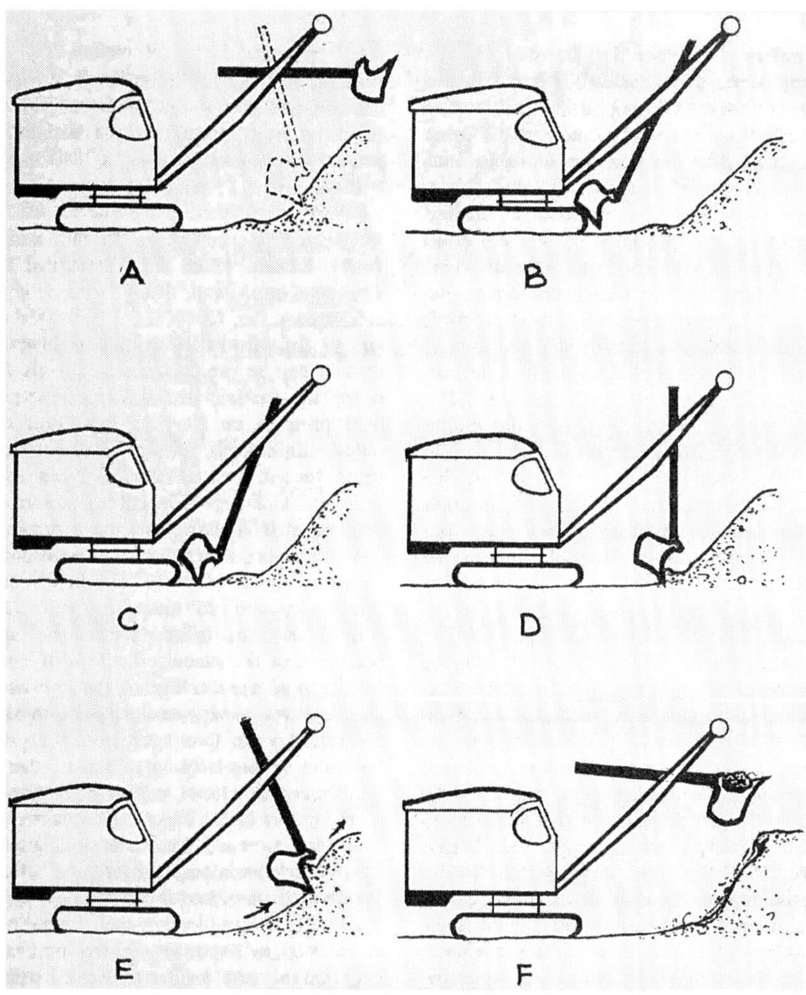

Fig. 5. Digging a bank

boost a boulder or stump to the side. The same criticisms apply as to swing-scraping.

The drive chains of a dipper shovel should be kept to the rear for both traveling and digging. In this position, power applied to the bull wheels is exerted directly on the lower tracks where they are in contact with the ground, and there is minimum rocking back and forth under digging thrust. In addition, they are less likely to get in trouble with stones and loose dirt.

Pull shovels and draglines work with the chains toward the digging.

Tangled Cable. If the hoist clutch and brake are released when the bucket is in the air, it will fall, unwinding the hoist cable from the drum rapidly. When the bucket strikes the ground it will stop pulling on the cable but the drum will continue to revolve with its own momentum, pushing the cable. This will result in the cable leaving the drum and looping itself all over the machinery, or the drum revolving inside the wraps of cable, thus loosening all of them back to the anchor, or in both. If the hoist clutch is engaged and the cable wound back onto the drum the wraps will cross each other and tangle. If a heavy load is picked up by the cable in this condition it will be strained, deformed, or crushed. In addition to having its life shortened it will then be difficult to spool onto the drum smoothly unless under enough load to pull its kinks straight as they come in.

The easiest way to straighten out this mess is to manipulate the bucket so that it will pull most of the cable away from the drum. It is first necessary to engage the hoist clutch and reel in the disordered cable, then hoist the bucket, dumping it first if loaded. The bucket may then be crowded all the way out and rested on the ground, dropped in a hole, or pulled back close to the tracks. There will now be only a few wraps of cable on the drum, which can be straightened out with the gloved hand, or gently with a screwdriver and a ball pean hammer. The engine clutch should be disconnected before handling the cable as an inequality in the hoist clutch might cause it to tighten the cable with a jerk even while disengaged. Leather palm gloves should be worn because even a preformed cable may have projecting broken wires that cut like needles.

If it is preferred to straighten the cable without raising the bucket, the engine clutch and the hoist clutch and brake should be disengaged and the cable pulled off the drum by hand until it is possible to straighten out the wraps still on it. The engine and hoist clutches should then be engaged with the engine idling, and the cable guided cautiously into correct position by hand as it is reeled in. It is sometimes easier if the cable is stripped from the drum by pulling it from the live side of the bucket sheave, as the boom point sheave will then line it up for rewinding.

If the cable has been loaded at all while tangled, it should be put under heavy tension while fully extended to straighten out kinks.

Tangling the cable in this manner is one of the principal bugaboos for beginners. It can be avoided almost entirely by riding the brakes, that is, by keeping a light pressure on the foot brakes at all times—not enough to stop the drum from being turned, but sufficient to stop free spinning. This is bad operation as it gives the shovel a constant drag to overcome, increases lining wear, and heats the drums. However, none of these drawbacks are serious enough to counterbalance the wasted time and labor, the loss of confidence, and the damage to the cable resulting from tangles. When the new operator develops good coordination with the controls, he can then train himself to release the brakes fully instead of partially.

The crowd retract cables should not tangle unless loose. They are straightened by prying with a screwdriver, and working the stick back and forth.

Cable Breaks. Shovel cables are subject to shock, heavy loads, sharp bending, rapid motion, exposure to weather, and drag cables may be exposed to friction with earth and rock. They may last a few hours or for years, but sooner or later they will break unless replaced. If the break occurs during work, time is lost until another cable can be obtained and installed. If the break occurs at the wrong time, it may cause injury or death to personnel and damage to property.

A cable that is abused, is too small for its work, or is of poor quality or defective may break suddenly before showing any

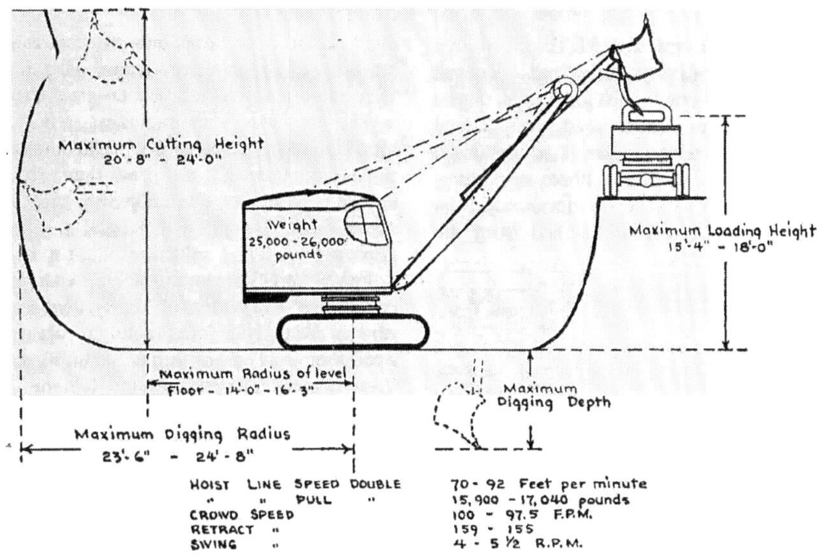

Fig. 6. Weight, range, and speed

signs of wear. Usually, however, a wire rope will not break until weakened and inspection will show thinned and broken wires on its surface. Such a cable may last a long time but it is good operation to change it during maintenance time in order to eliminate work stoppage and possible damage when it fails.

Cables often break one strand at a time, giving the operator warning before parting completely. However, the fact that fifty cables may break in this manner is no proof that the fifty-first will not snap suddenly, either under load or reaction from load.

Safety Precautions. An operator should lower the bucket to the ground before leaving the machine as the brakes may lose their grip as they cool and let the bucket down, gradually or with a rush. There is also the possibility of a would-be operator accidentally releasing the hoist brake while an admiring friend stands under the bucket.

Workmen and spectators should always keep beyond the furthest reach of the shovel, if possible, but they seldom do. A point to remember when in danger of being hit by the bucket is that a shovel is weak on the infighting, and that it may be safest to run toward it where the slower speed of its motion makes it easier to dodge and less damaging if it does connect.

Booming Up and Down. The boom is normally held in a fixed position during digging. However, lowering the boom enables the bucket to dig deeper and further away and raising it permits a higher dump point. If desired, the boom latch may be locked out of engagement, the boom lowered to dig, raised during the swing to dump, and lowered again on the swing back. This uses extra power to lift the boom and may slow the digging cycle, but often simplifies the work. It is only possible with a live boom.

Booms should always be lowered slowly and stopped with the brake, not the ratchet.

Reach and Speed Data. Figure 6 indicates the working range and speed of half yard dipper stick shovels. These are taken from the published specifications of several standard makes. For simplicity, the differences caused by boom angle have been omitted, and each distance is the maximum that can be reached from the most favorable angle. Live boom machines can operate according to these figures for depth of digging, reach, and dumping height, by raising or lowering the boom during the swing. Other machines can be adjusted for maximum performance for one or two, but not for all of the positions shown.

TYPICAL OUTPUT TABLE				
	PER HOUR	SHIFT 8 HOURS	WEEK 40 HOURS	YEAR 2,000 HOURS
3/8 Yard	42	336	1,680	84,000
1/2 Yard	63	504	2,520	126,000
3/4 Yard	92	736	3,680	184,000
1 Yard	122	976	4,880	244,000
1 1/4 Yard	151	1,208	6,040	302,000
1 1/2 Yard	176	1,408	7,040	352,000
1 3/4 Yard	197	1,576	7,880	394,000
2 Yard	223	1,784	8,920	446,000
2 1/2 Yard	260	2,080	10,400	520,000

Based on average digging and on operating efficiency of 84% (average 10 min. delay per hour)

Fig. 7. Dipper shovel output

A shovel does not work efficiently at its maximum reach as it loses in lift, in ability to fill the bucket in a short sweep, in tooth angle, and in accuracy of control. Good digging practice is to keep the shovel close to the work and trucks down at its level for easy, accurate loading. Special situations often arise, however, where every inch of reach must be utilized.

Swing speeds are quite variable, ranging from four to five and a half revolutions per minute. This difference is not very important, as the gain in speed usually means a loss of power. A slow swinging shovel may swing from digging to close dumping position more quickly than a fast swinging one of the same power because of more rapid acceleration, but will be slower on a long swing.

Special dipper shovels, called stripping shovels, are made with booms and sticks proportionately much longer than those shown. Counterweight and power may be increased, or bucket size reduced to compensate.

Production. The production of a shovel cannot be predicted accurately, because of the many variables in the machines, operators, and soil. The table in Figure 7 represents average output loading a steady stream of trucks, in medium digging, with a ninety degree swing, figuring fifty working minutes to the hour. It should not be used as a basis for estimates, as many jobs apparently having these conditions will do well to attain half this yardage.

Production will be increased if the swing angle is reduced, if the soil is easier to dig, or heaps higher on the bucket, or average time loss is reduced.

Production will be reduced if the shovel must wait for trucks to get in position, if the bank is so low as to require frequent moves, if loading must be done at a higher level, if the digging is obstructed by roots, stumps, or boulders, or is too hard to yield a full bucket with each pass, and if the swing is longer than ninety degrees.

As digging becomes hard, the output of small machines drops faster than that of larger ones. In cramped spaces, with limited working room, output of large shovels falls off more than that of the small ones.

Non-caving or slow-caving banks should not be higher than the shipper shaft, nor less than half that height, for best results. If the soil is hard, the higher bank may be advantageous as it gives longer contact for the bucket, so that it can fill from a thinner slice.

Banks that slide freely as the toe is

dug away give best loading conditions and are limited in height only by danger of heavy slides.

Side casting will increase output from twenty to one hundred percent over truck loading, as output can be continuous, without pauses for accurate spotting. Large trucks are more quickly loaded than small ones if body walls are not inconveniently high.

Operating Suggestions. The following is a list of suggestions for efficient dipper stick operation:

1. Keep bucket teeth sharp and built up to proper size.
2. Don't crowd bucket so deep into the bank that it slows the hoist.
3. Pull bucket out of the bank as soon as it is full.
4. Keep close to the bank unless it is likely to slide.
5. If bank is very hard, or material needs mixing, make passes in it with bucket door open.
6. Place shovel so that the drive chains are away from the digging.
7. Spot trucks as close to digging as safety permits to save swing time.
8. Spot trucks on both sides if possible to save waiting time while trucks move.
9. Spot trucks so that they line up with arc of swing, or back them directly toward the shovel.
10. Don't swing over truck cab if you can avoid it.
11. Move up while waiting for trucks.
12. In mud, load driest material in bottom of truck, sticky stuff on top, for easier dumping.
13. Don't let tracks sink deep in ground—bury mud with dry fill or use poles or platforms.
14. Work smoothly—slamming and jerking are hard on machine and operator.

DIGGING WITH A PULL SHOVEL

The dipper shovel is primarily designed to dig at its own level. The pull shovel, Figure 8, also known as a hoe, backhoe, dragshovel, and ditching shovel is at its best digging below the level on which it stands.

The stick is pivoted on the end of the boom. The hoist line pulls on the stick sheaves on its top or back, and the drag cable pulls the bucket. The functions of hoist and drag are inter-related in a very complex manner.

If the hoist clutch is engaged the stick sheaves will be pulled toward the top of the jack boom, the stick will pivot on the boom end, and the bucket will move outward as in Figure 9 (A). If the drag brake is held, or the bucket encounters resistance so that it cannot move out, the stick will be held against pivoting, and stick and boom will be pulled toward the jack boom as a unit. The boom cannot move in a straight line but pivots on its foot pins so as to rise as it comes back as in (B).

If the hoist brake is locked and the drag slipped, the bucket will move out and somewhat down, (C).

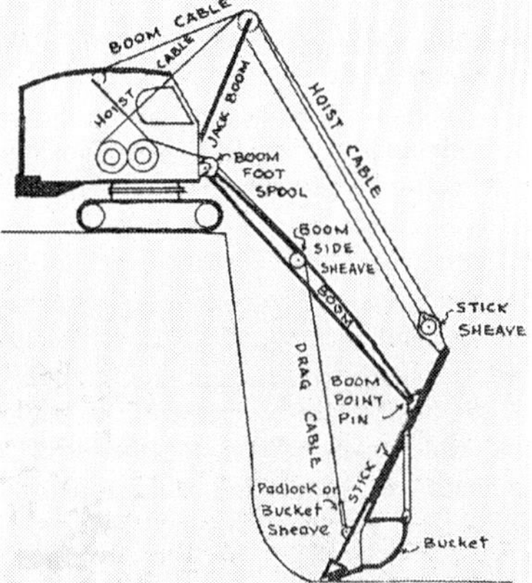

Fig. 8. Hoe reeving

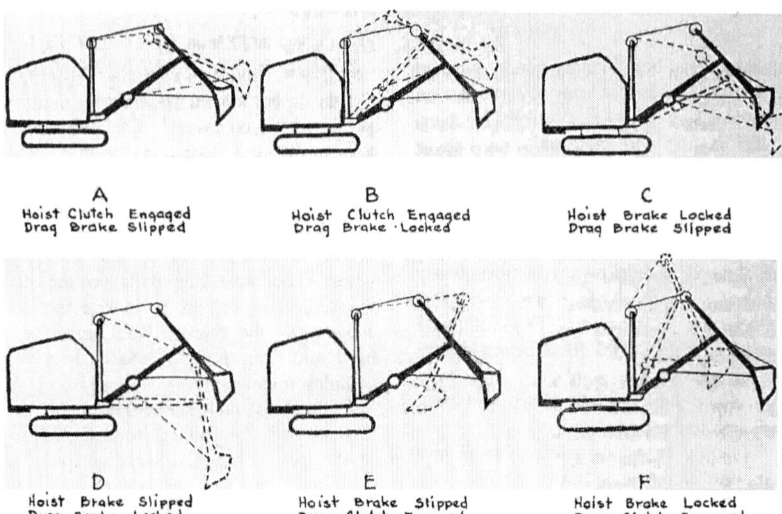

Fig. 9. Hoe bucket positions

If the hoist brake is slipped and the drag brake is held, the boom and stick will swing down in an arc, as in (D). If the drag clutch is engaged while slipping the hoist brake, the bucket will move in horizontally, as in (E).

If the hoist brake is locked and the drag clutch engaged, the bucket will move in, but the stick, pivoting on the boom, will move the stick sheave out, tightening the hoist cable and causing boom and bucket to rise, as in (F).

Starting Cut. The bucket can be moved out horizontally from any raised position by releasing or slipping the drag brake so that the bucket swings out, and engaging the hoist clutch enough to prevent hitting the ground, as in Figure 10 (A). When it is almost fully extended or is at the desired distance, the drag brake is applied and the hoist drum is released. The boom, stick, and bucket will fall together striking the teeth on the ground with a pickaxe effect, (B). For soft digging the drop is cushioned with the hoist brake. In penetrating a hard surface the brake is applied just as or after it strikes, in time to prevent cable from spinning off the drum.

If the drag clutch is then engaged and the hoist released, the bucket will move toward the shovel with its teeth in digging position. If the ground is normal the bucket will rapidly cut its way down until full, then resist any further pull.

If in the (B) position the drag clutch is engaged and the hoist brake locked, the bucket will move in causing the stick to pivot on the boom end. The stick sheave, held from pivoting forward by the hoist cable, acts as a fulcrum and the stick, now a second class lever, forces the boom point toward the shovel. The boom pivots on the foot pins, rises in an arc pulling the bucket up with it, and causes it to lose contact with the ground.

Since releasing the hoist brake will cause the bucket to dig too deeply, and locking the brake will cause the bucket to rise above its digging, it is evident that keeping the bucket at the desired depth while being pulled in may be accomplished by either partial or intermittent use of the hoist brake. A smooth, sensitive brake can be partially engaged; a jumpy one is applied in a series of light jerks.

The bucket first acts somewhat as a dozer blade and pushes dirt (C), then curves under it and fills, and as it passes under the boom point the teeth start to point upward (D), and come out of the ground. The hoist clutch is engaged and drag continued until the bucket sheave guard nearly touches the boom, (E), when the drag clutch is released and the drag brake is applied with hoist continuing during the swing.

Dumping. When dumping position is reached, usually a 35° to 90° swing for ditching, or 45° to 180° for loading trucks, the drag brake is released and the bucket swings outward, being prevented from grounding by holding the hoist brake, or engaging the hoist clutch. Material on the teeth and lip falls first, then that in the body of the bucket, and last, dirt that is lightly stuck to the sides and back. More adhesive stuff may be shaken out by applying and releasing the drag brake in jerks during the dump.

The dump may be spread out over a distance of some twelve feet but is likely to be concentrated in a shorter distance. Loose, dry material dumps near the machine, and compact, sticky dirt at the furthest point where the bucket bottom is vertical or overhanging. The whole contents of the bucket may be dumped in the nearer area by only partially releasing the drag brake, so as to keep the bucket over one spot as the hoist pulls it up and overturns it, position (H).

After dumping, the bucket is dropped again at the beginning of the ditch and pulled in digging another full length slice, and this maneuver is repeated until the desired depth is obtained. The digging may also be done by cutting the bucket in deeply after the first penetration, and lifting when filled, leaving the remainder of the slice for the next bite. If the spoil is

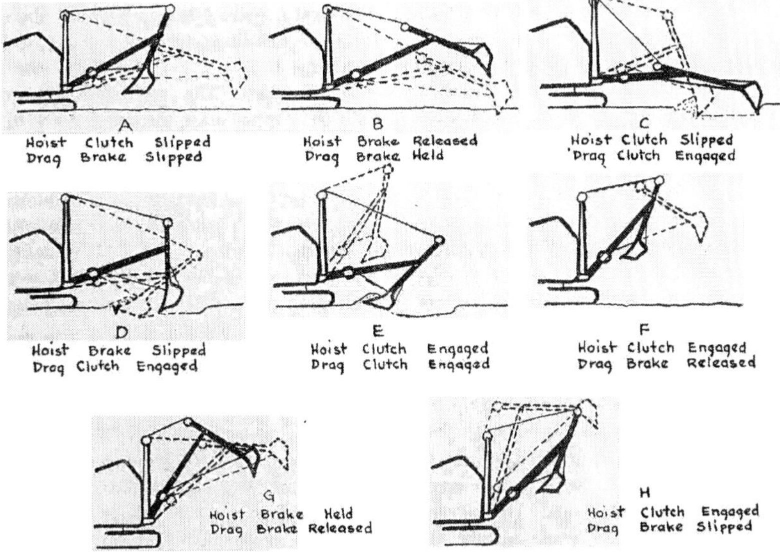

Fig. 10. Digging and dumping

to be dumped immediately alongside the ditch, spillage is unimportant, and the bucket can be hoisted and dumped without pulling it close to the boom.

Push Back. If a large amount of earth is to be dug, it is good practice to push the spoil pile back while dumping on it. This may be done by releasing the drag brake before hoisting far enough to clear the pile. The loaded bucket swings down and out by gravity, striking a blow against the heap which, if correctly aimed, will knock the top off. The momentum of the bucket plus the weight of the boom against the stick and the continuing hoist pull give a follow-through which should push the dirt out beyond dumping range, making room for the bucket contents in the pile. This operation should be started while the pile is low and the blow has maximum force.

This technique may also be used to smooth over loose ground on which the shovel must walk, to boost boulders out of the way, and to dispose of piles left at the near end of trenches. The novice should use it with caution as mistakes in aim or timing will cause cable tangles.

Loading Trucks. The most convenient position for a truck to take to be loaded by a backhoe is very close to it to reduce spilling, and backed toward it to allow dumping in the full length of the body. The danger in this position is that a broken drag cable or other accident may permit the bucket to swing outward, taking the truck cab with it. For this reason trucks are often loaded from the side.

A pull shovel loads trucks more easily if it is on a higher level, so it is good practice to have the trucks in the pit or to build a dirt platform for the shovel. This last may be done by utilizing the spill at the rear of the truck or by dumping earth where the shovel is to walk. The advantage is that a low boom angle permits the bucket to be held in position to retain a full load at some distance from the shovel, whereas a high boom puts the bucket in a partially dumped position even when just forward of the tracks.

Boom Fall-Back. The pull shovel is particularly vulnerable to having the boom fall back. Either the hoist or drag if engaged too long will pull the boom back on the jack boom, and both of them on top of the cab. Also, if the shovel does not stand level a boom and bucket in a position which is safe on the low side may fall back on the cab when it swings to face uphill. A warning of this is a slackening of the drag cable, a cure if it is not too late is to release both brakes and let the boom down and the bucket out until balance is restored.

A structural safety measure is to hinge a piece of pipe to the A frame, and another to the jack boom, of such sizes that one will slide inside the other. A stop pin may be put in either to prevent them from telescoping together far enough to allow the jack boom to touch the cab, without interfering with normal use. If the A frame is a folding type, it might require bracing.

Applications. The pull shovel shares with the dipper stick the advantages of a rigidly attached bucket, with resultant control of digging, and ability to hold the bucket teeth to their work. In addition it is able to work below the grade on which it stands, keeping itself and sometimes attendant trucks out of the mud, rock, and ramp difficulties which often hamper digging from the floor.

The pull shovel is adept at stripping top soil, making shallow cuts and removing windrows without losing material at the sides, at removing overburden from rock prior to quarrying, at scraping down high banks, and particularly at its specialty of digging small cellars and ditches to twelve or fourteen feet in depth.

However, it is sloppy and inefficient at

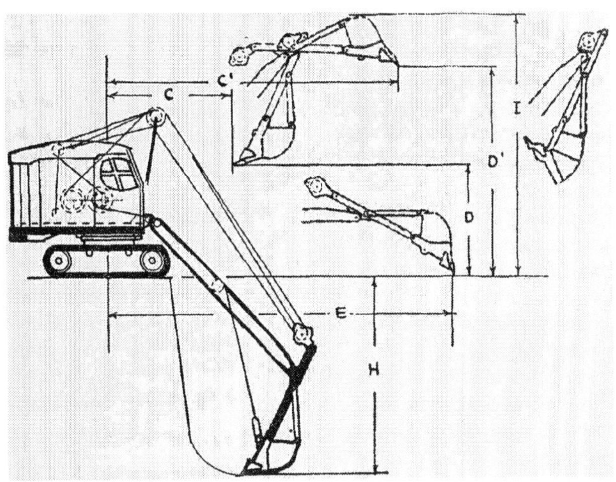

	MIN.	MAX.
BOOM LENGTH	14'-0"	17'-0"
STICK	6'-11"	7'-9"
H DEPTH	12'-0"	16'-0"
C RADIUS - BEGIN DUMP	9'-0"	9'-0"
C' RADIUS - END DUMP	19'-6"	22'-6"
D CLEARANCE - BEGIN DUMP	8'-9"	9'-6"
D' CLEARANCE - END DUMP	14'0	16'4"
E DIGGING RADIUS DUMP	22'-9"	27'-9"
I MAX. HEIGHT	19'-6"	22'-0"

	MIN.	MAX.	
HOIST LINE SPEED - SINGLE	180	184	F.P.M
HOIST LINE PULL - SINGLE	8,000	8,500	lbs.
HOIST LINE NO. PARTS	2	4	
DIGGING LINE SPEED - 2 PART	62	90	F.P.M.
DIGGING LINE PULL - 2 PART	12,800	32,000	
WEIGHT	25,000	30,700	

Fig. 11. Hoe range and speed

loading trucks and has a slow digging cycle. The sloppiness is due to material falling off the teeth as the bucket is lifted toward dumping position, and the inefficiency results from the maneuvering necessary to complete a dump within the length of a truck body. The slow cycle is partly because of the carrying of the extra weight of a boom in digging motions and largely because of the necessity of pulling the bucket close into the boom before hoisting, in order to retain the load during the hoist.

A bucket which can be rotated from digging to dumping position greatly increases pull shovel efficiency. At present these are generally available only on hydraulic backhoes.

Range and Speed. Figure 11 gives range data for half yard hoes. The boom is always live and the angle of the jack boom has little effect on the digging, so that the performance shown can be attained by either live or dead boom units.

The bucket fills best well below the tracks, but it loses penetration near maximum depth. This is because the teeth cannot strike the ground at a steep enough angle, and the direction of pull is more upward than inward.

Hoes are not usually made with bucket capacities greater than two yards.

Production. There are so many variables in hoe operation that a table of output would be of little use. Yardage may run from fifty to eighty percent of that moved by a dipper stick, under conditions favorable to them both.

Suggestions for operation. The following suggestions for pull shovel operation may be of use:

1. Keep bucket teeth sharp and built up to proper size.
2. Line shovel up to cut outer edges of excavation in a straight line.
3. In deep digging, keep a fairly straight face and keep shovel back from it as far as possiple. Don't work out on a "peninsula" which may cave.
4. Keep knocking the spoil pile back as you dump on it. You may need the space.
5. Don't dig yourself into a trap—the spoil pile may slump across your way out.
6. Don't let the drag slide the shovel into trouble when the bucket hooks something solid.
7. Don't work with the drag cable in the dirt. If a pile builds up in front of the tracks, back away and knock it off.
8. Try to keep the shovel higher than trucks it is loading.
9. Don't pull boom back on top of shovel with drag or hoist.
10. Keep boom low while walking, or while swinging uphill.
11. Don't walk, or swing uphill, when drag cable is slack.

OPERATING A DRAGLINE

A dragline attachment is shown in Figure 12. It has a long light crane boom, with a fairlead set at its foot, and a bucket attached to the machine only by cables.

Reeving. The dump cable runs from the top of the bucket arch over the dump sheave and forward to the drag yoke.

The boom line is a standard four-part

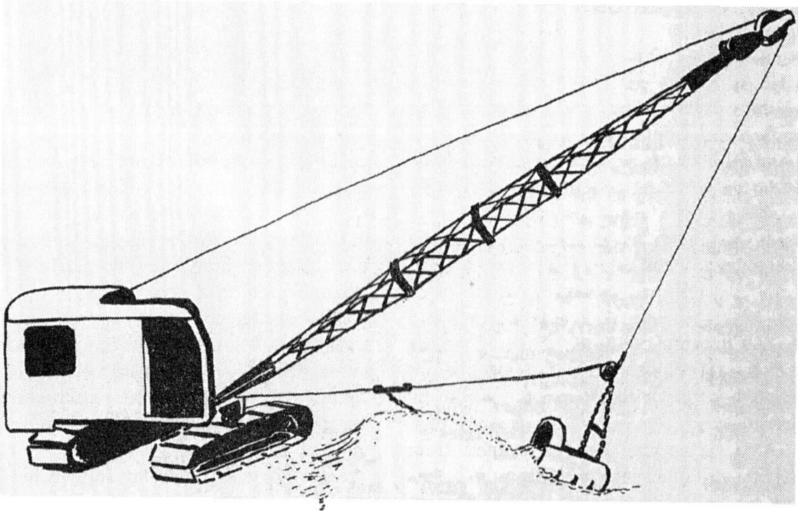

Fig. 12A. Dragline shovel

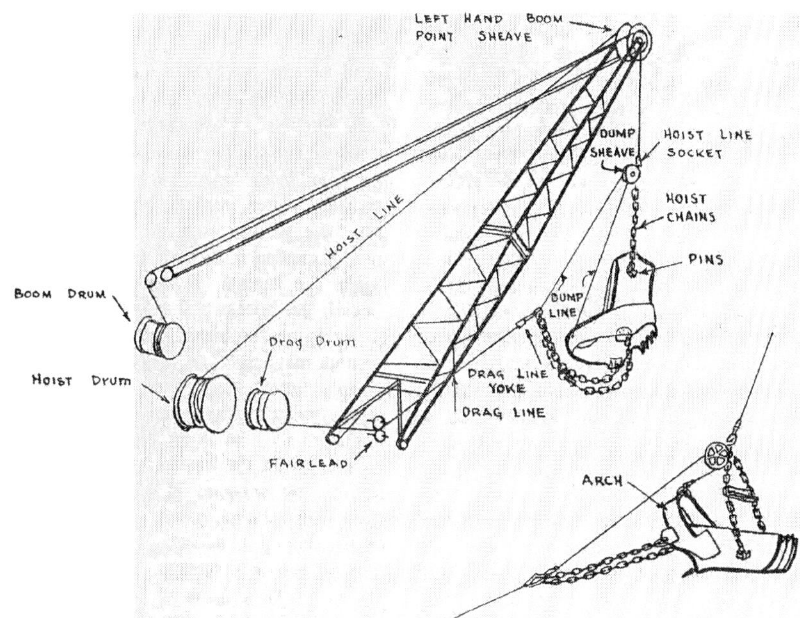

Fig. 12B. Dragline attachment

rigging, similar to the dipper boom support, except that a longer cable is needed. The hoist line runs from the hoist drum over a large boom point sheave and down to the dump sheave case. The drag cable runs from the drag (digging) drum through the fairlead to the drag yoke.

Bucket Action. If the bucket is lifted with the hoist while the drag cable is slack, it will hang in fully dumped position. If tension is then put on the drag cable, it will pull on the dump cable before the slack is out of the drag chains. The dump cable will pull the front of the bucket up, toward the dump sheave, 12B lower. Releasing the drag cable will allow the dump cable to run back over the sheave, and the bucket will return to dump position, pivoting on the hoist chain pins.

If the bucket is then lowered to the ground, it will turn to a horizontal position, or will rest on its teeth and arch, depending on its balance. A pull on the drag cable will now tip the bucket forward or backward onto its teeth and the teeth and lip will dig in as it is dragged toward the shovel. If the pull is continued with the hoist line slack, the bucket will cut to a depth determined by its weight, the angle and sharpness of its teeth, and the resistance of the soil. If so deep a cut is not wanted some tension is put on the hoist line, raising the bucket slightly. If the dump cable is long the bucket will be raised in the rear by the hoist chains. A short dump line will cause an upward pull on the arch, raising the front of the bucket as much or more than the rear. In either case the depth of cut is reduced. A further pull on the hoist will raise the bucket clear of the ground.

Whether the bucket will remain in the carrying position, or partially dump while being raised, depends on opposing forces acting through the dump cable. The weight of the bucket front pulls down on the arch end of the cable and the tension between the hoist and drag cables tends to stretch its drag-yoke-to-dump-sheave section and pull the bucket up. In effect, the dump cable must pinch the other two cables together in order to obtain slack to drop the front of the bucket. This pinching requires comparatively little force when the angle between the cables is small and becomes increasingly difficult as the angle flattens out. Also, if the dump cable is long it will not have to pull as strongly on the two cables to obtain slack as if it is short. The action of the bucket will therefore depend on the angle between the hoist and drag cables, the length of the dump cable, and the weight and distribution of the bucket load.

A wide angle between hoist and drag cables can be had when picking the bucket out of the soil, either by pulling the bucket close to the shovel, or by keeping the boom at a low angle. Bringing the bucket all the way in usually wastes time and causes wear on bucket and chains which might be avoided if it were picked up as soon as full. A low boom has a tendency to tip the shovel when heavily loaded, and often cannot be used because of obstructions or height of dump. A short dump cable makes it difficult to dump except directly under the boom point. If a live boom is used it is possible to dig low and dump high, but this takes extra time and work.

The technique used will depend on the job and on the operator's preference. There is generally at least one good method of handling any situation.

Digging. In an ordinary dragline digging cycle the bucket is not thrown or cast. It is lowered into the pit with both lines taut, the hoist brake being almost wholly released, then re-applied smoothly as the bucket is about to strike the ground, and the drag brake is released enough to allow the bucket to drop straight instead of following an arc centering on the fairlead.

When the bucket rests on the ground the hoist cable is slackened slightly and the drag clutch engaged. The drag cable pulls the bucket, with the teeth digging in and cutting a slice of dirt which piles inside the bucket. If the hoist brake is locked, the bucket will move up in an arc centering on the boom point, and on level ground may pivot so the teeth dig more sharply but no longer have the full weight of the bucket to force them in.

Ordinarily, the hoist brake is released enough to let the bucket cut level or follow the pit contour. If the pit slopes up toward the shovel, which is the most favorable digging condition, it may be necessary to partially engage the hoist clutch to avoid digging in, or to prevent the hoist line from becoming too slack and allowing the chains and dump sheave to slump into the bucket.

Hoisting. When the bucket is filled the hoist clutch is fully engaged and the bucket is lifted clear of the ground. If the bucket has a tendency to dump, the drag may be left engaged until the angle between drag and hoist cables is sufficiently wide to hold it. The drag clutch is then released and the hoist is continued, with the drag brake applied just enough to allow the hoist to pull the bucket forward and upward under the boom, without slackening the lines enough to dump it. If the drag brake is applied too tightly the bucket may hit the boom.

The swing is started as soon as the bucket is clear of the ground. This is the period of heaviest load in the dragline cycle, as the machine is simultaneously lifting a load, pulling against the drag brake with the hoist clutch, and swinging. If overloads are picked up in the bucket they may so slow the line and swing speeds that less dirt will be moved per hour than if smaller bites are taken.

Dumping. Hoisting is discontinued as soon as the bucket is high enough to be dumped. When the swing is completed, the drag brake is released, partially or wholly, and the bucket swings out and dumps. A long dump cable will permit normal dumping inside the boom point, a short one under it only. Raising the boom will bring the dump point closer, lowering it puts it further away.

It is poor operation to raise the bucket higher than necessary before dumping. Clearance should be allowed so as not to strike trucks or other receptacles, but piles can be barely cleared. This saves time on a short swing and fuel and wear and tear under any circumstances.

Casting. If the bucket is to be thrown or cast it is pulled close in by the drag cable during the swing to the pit, with the hoist line held at a length that will keep the bucket above any obstructions during the cast, as in Figure 13 (A). When the swing is completed, the drag brake and clutch are released and the bucket swings outward like a pendulum. When it is just short of the furthest point of this swing, the hoist brake is released and the bucket falls downward and outward (B). It is checked by gradual application of both brakes just before hitting the ground, as otherwise it might be damaged, overturned, or tangled in its chains. It should then rest on the ground in the same position as if lowered and responds to drag pull similarly, except that if the pit floor is level, the hoist cable slackens as the bucket approaches the boom point and tightens again after it has passed under it.

A swing throw may also be used. During the swing from dump to pit, the drag cable is left slack and the hoist held so that the bucket is high enough to clear obstructions. The centrifugal force of the swing will pull the bucket outward, and

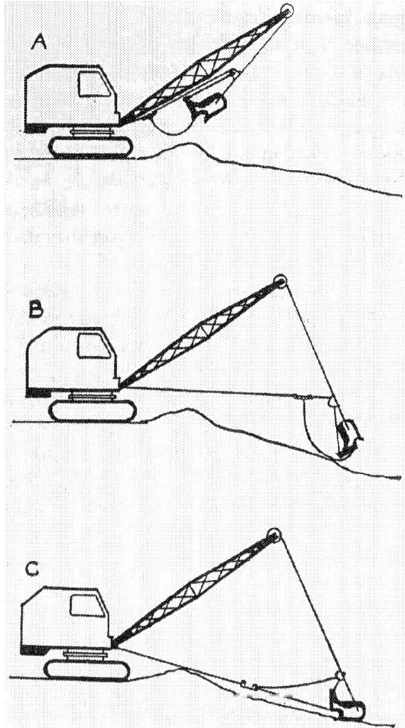

Fig. 13. Casting the bucket

the hoist brake should be released at the proper point so that the bucket will land where it is wanted. The swing should be checked as soon as the hoist brake is released, and hoist and drag brakes applied gently as or just before the bucket strikes. This swing throw requires much more expert operation than the other type and may damage the shovel seriously if improperly done.

Another technique is to make a pendulum throw while the shovel is swinging so that the centrifugal force adds to the outward sweep of the bucket.

Throwing to dump is a similar process, but usually the height of the piles prevents the use of a long enough hoist line to obtain much distance. The weight of the load, however, causes the bucket to cast further than it could on the same length line if empty. A combination pendulum and swing throw is most effective, particularly if the shovel is dumping at 180° from the pit and can revolve in a full circle so that it can dump without stopping.

If the load sticks in a bucket which is thrown to dump, it may overturn the shovel.

The distance the bucket can be thrown is affected by the skill of the operator, the length and angle of the boom, the weight of the bucket, the depth of the pit, and even by the wind. Casting onto a surface level with the tracks, the teeth ought to reach further than the boom point would if lowered to a horizontal position. Much greater distance is attained when casting into a pit.

Manufacturers are wary about recommending or discussing throwing of buckets, because a careless operator may thus bang a good bucket into scrap iron in a short time, and the swing throw has possibilities for wrecking the boom as well. Beginners will do well to thoroughly master ordinary digging before practicing throws, particularly if they have trouble with tangled cables.

Throwing the bucket usually slows the digging cycle by several seconds, and reduces accuracy of work, so that digging should be done inside the boom point when practicable.

Tangles. If the drag cable becomes tangled it is good procedure to throw the empty bucket, as often the whole tangle can be unwound in this way, and at worst very few wraps will be left on the drum to be straightened. If the drag cable becomes bent or kinked so that it does not spool on smoothly, the bucket should be dragged in all the way a few times with a load to straighten it. If the bucket is swinging in the air the drag should be wound in while it is swinging out and holding the cable under tension, and held while it is swinging in and the cable is slack. If the cable is running smoothly, this precaution need not be taken.

If the hoist cable becomes tangled the bucket should be rested on the ground, the hoist brake released, and the machine swung. When the hoist cable is unwound past the tangles or to the anchor, the hoist clutch should be engaged and the swing clutch disengaged. The cable should then reel back properly, unless crushed or kinked, in which case tension should be put on it as it is wound in, by partially engaging the swing clutch to keep the boom from swinging back easily over the bucket.

The cables should not be allowed to run all the way out in a throw, for if the momentum of the bucket is stopped abruptly by the cable anchor the anchor may tear out of the drum. Manufacturers often supply and specify a drag cable which is so short that it is unsafe for casting.

Novice Operators. To the novice, the dragline is apt to seem a very loose, rough, and contrary machine. Special provisions have to be made in every movement to keep the bucket from jerking and swinging, but these soon become automatic and very precise control can be obtained eventually. The beginner will obtain the best results by keeping the bucket fairly close in, except when actually digging or dumping, by making slow starts and slow stops when swinging, and dragging the brakes slightly to avoid spinning out the cables.

Bucket Wander. As the bucket is pulled in, it is liable to be deflected by irregularities and find a path considerably to one side of a direct line. Also, the shovel may swing by gravity during the haul-in if it is not standing level. In either case the drag cable will not be directly under the boom. The fairlead sheaves will put it in correct alignment for the drum, but if the angle is sharper than the fairlead pivot can meet, the cable will be dragged across the fairlead guard plate and will wear the cable, and wear or possibly tear off the guard. The shovel boom should be kept in line with the cable by the use of the swing lever.

Boom Twist. As a loaded bucket is lifted the boom will tend to swing over it. However, if the boom is held to one side by the swing clutches during hoisting, or if a heavy, oversize load such as a stump is partially lifted, then dragged along the ground by the swing, a powerful twisting force is applied to the boom.

In order to have a long reach without prohibitive weight, a dragline boom is of light skeleton construction that is not intended to withstand heavy side and twisting strains. Sometimes a boom so strained will collapse, but more often will twist slightly, bending some of the cross braces particularly on the lower side. Once twisted, normal loads may increase the damage and failure will follow if the boom is not straightened.

A good ironworker is needed to properly repair such a boom, but if the bent members are angle irons it may be kept in service quite a while by straightening them with a jack. A stout plank is placed inside the boom across the corner angles as a support for an automobile jack that can push the pieces straight.

This type of repair should not be attempted on tubular members as they lose their strength if flattened.

Applications. The dragline does not have the positive digging force of the dipper and the pull shovel, as the bucket is not weighted nor held in alignment by rigid structures, and can therefore bounce, tip over forward, or drift sideward when it encounters hard material. This weakness increases with depth, and is particularly noticeable in small size machines, as the weight and bulk of a large bucket are sufficient to give considerable stability and penetration.

The dragline experiences its greatest difficulty in cutting down and is able to continue a deep cut opened by other machines or by blasting in much harder material than it could dig from the surface.

The outstanding advantage of a dragline over most other rigs is long reach for both digging and dumping, plus the ability to dig below the tracks. It has the further good point of a high cycle speed, being second only to the dipper stick in this regard. It will be preferred to the dipper stick for truck loading where the earth is not too tough, and where the original grade is better than the new, because of water, mud rock outcrops, steep ramps, or other problems within the excavation. Because of its greater reach and output, it will be preferred to the pull shovel in any situation where it is capable of digging the soil effectively, where precisely cut vertical sides are not required, and where there is room for it to swing.

The dragline is the only practical attachment for extensive digging in mud, as its reach enables it to handle a wide area from a single stand and the sliding motion of the bucket avoids trouble with suction. It is also the best machine for many stripping operations in which the spoil is high-piled away from the pit, but is rivaled by long boomed dipper stripping shovels if the spoil is to be moved across a narrow pit as in some strip mining.

Draglines with skillful operators do an excellent job of topsoil stripping, grading, and spreading piles of earth, but except under wet conditions they do not usually do as well as bulldozers of comparable value.

Working Data. Figure 14 gives working diagrams and data for half-yard draglines with thirty foot booms. Shorter or longer booms may be mounted. A long boom in-

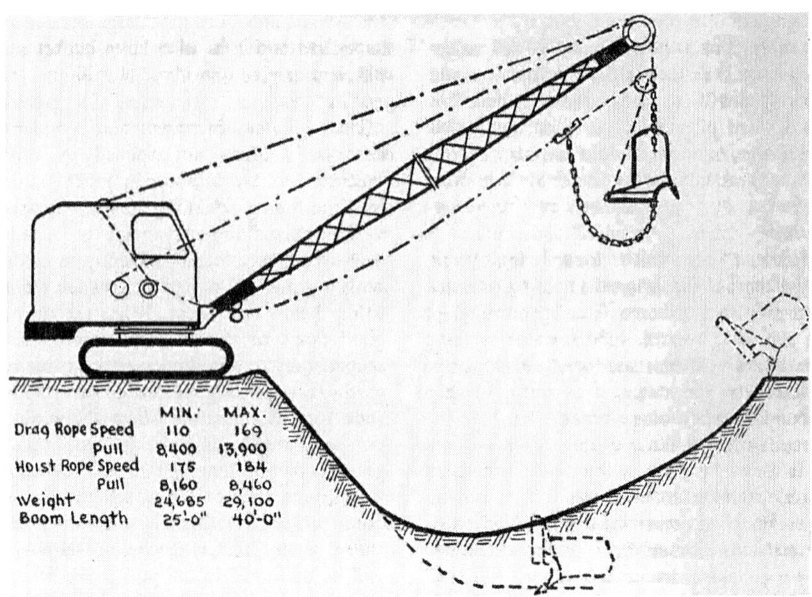

Fig. 14. Dragline speed and pull

creases reach and depth, but may require additional counterweight, reduces the safe load, slows the swing, and gets into difficulties with obstructions.

As in previous tabulations, the data are averaged from that published for several different makes.

Production. In easy digging, dragline production should be between sixty and ninety percent of that of a dipper shovel of the same size if trucks are loaded on the same level. If trucks are in the pit output may range from seventy to a hundred percent.

In hard digging, dragline yardage falls off much more rapidly than shovel yardage because of weaker penetration. High banks also slow them when loading on their own level or side casting, because of the extra time required for hoisting.

In side casting production on a volume-distance basis may be much higher than that of a dipper.

Operation Suggestions. The following are suggestions for dragline operation:

1. Keep bucket teeth sharp and built up to proper size.
2. Keep dump cable short so load can be picked up well out from the shovel.
3. Dig in layers, not ditches.
4. Keep digging surface sloped up toward shovel.
5. Don't drag in so far that dirt or rocks pile up in a ridge.
6. Keep drag cable from working in dirt.
7. Pick up bucket as soon as it is full.
8. Do not pull drag yoke into fairlead.
9. Inspect bucket chains frequently, especially at ends, and have them built up or replaced if worn thin.
10. Inspect fairlead frequently—worn bushings or spacers may let sheaves wear and cut cable.
11. Salvage short pieces of hoist or drag cable for dump cable. If dump an-

chors are too small, hammer the cable flat or change the anchors.

12. Do not guide bucket by swinging the boom while digging—you may twist the boom.
13. Do not swing until bucket or load is clear of the ground—you may twist the boom.
14. For heavy loads use high boom and swing slowly.
15. If the shovel starts to tip dangerously, drop the bucket. Don't try to dump it as its outward swing increases the tip.
16. Don't pick up overloads in bucket if they slow the hoist and swing, or if they tip the shovel.
17. Don't slap bucket against boom while hoisting.
18. Work with machine level. If you can't, load trucks at either the top or bottom of the swing. Loading is sloppy at the side as the counterweight swings the boom uphill as the load is dropped.
19. Don't work with a cable that is cross wound on the drum.
20. Don't travel uphill with a high boom.

RUNNING A CLAMSHELL

The clamshell front is a jack of all trades. It can do most of the jobs the other rigs will do, although less efficiently, and in addition has its own specialties of digging deep, narrow, straight-sided excavations, and neat rehandling of materials.

Figure 15 shows this attachment. It consists of a boom similar to that used for the dragline, with two boom line sheaves and two operating sheaves at the point, and a bucket hung from the operating sheaves by two cables.

Digging. In digging, the bucket is placed over the work by swinging the boom, and either moving the shovel or raising or lowering the boom to obtain correct distance. The digging (drag, closing, or crowd) brake is released, causing the bucket to open, and the hoist brake is then released, allowing the bucket to contact the ground

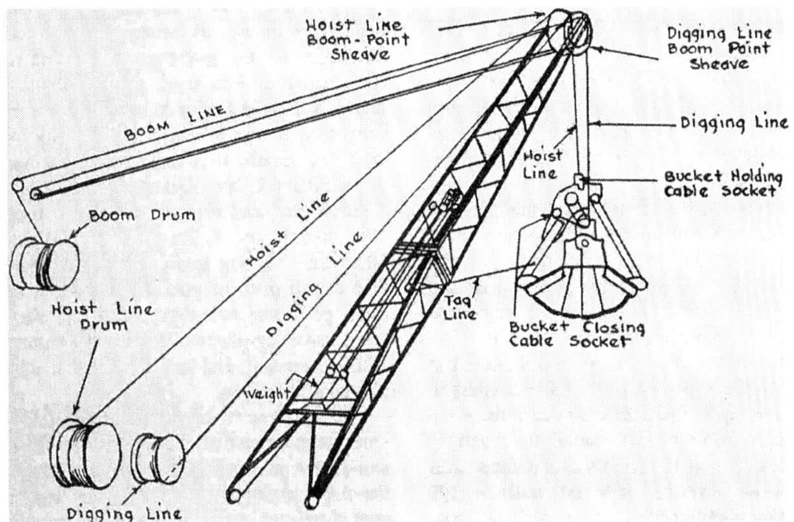

Fig. 15. Clamshell reeving

on its teeth. If the ground is soft, the hoist brake will be only partially released, or re-applied just before the bucket lands. If it is hard the bucket will be allowed to fall freely, so that its weight will drive it into the ground for a good bite. In either case both brakes must be applied as soon as it has hit to avoid unspooling of the cable.

The digging clutch is then engaged to pull the jaws together. They first push the dirt inward, then curve and close under it. If the material is not too resistant and the bucket has proper weight a full or heaping load will be gathered. The digging line will lift the bucket as soon as the jaws are tightly closed. The hoist clutch should then be engaged so that the hoist (holding) cable will not become slack while the bucket is being raised.

Dumping. The swing is started as soon as the bucket is clear of obstructions. The distance of the bucket from the machine may have to be adjusted by raising or lowering the boom if precise dumping is being done. This, of course, is practical only if the shovel is equipped with a live boom. When the bucket is properly positioned, the digging brake is released, the jaws open by their own weight and that of the load, and the earth is dumped. The bucket is swung back to the pit in the open position and lowered or dropped for another bite.

Centrifugal force of the swing puts the bucket out beyond the boom point, offering some increase of reach for both digging and dumping. Stopping the swing will allow the bucket to swing inside the boom point for close digging.

Chopping. If the bucket does not fill at the first closing it may be opened, hoisted, and dropped again, in which case the earth scraped together the second time will be added to that loosened the first time. This process may be repeated until a full bucket is obtained.

Signals. Clamshell work is often done at such depths that the operator cannot see the digging point. In this case it is best to have a man placed so that he can see the work and signal the operator. This is essential if there are men working in the pit where they might be hit by the bucket.

Hand Labor. If the bucket will not dig satisfactorily, either because it is too light in construction or slides off slopes, hand labor or explosives may be employed to loosen the dirt and the clamshell used to lift it out afterward.

Deep Digging. In deep ditches or shafts, it is difficult to keep the walls perpendicular, particularly if the earth is stony, as with each bite the bucket is edged a little away from the wall, thus causing the pit to grow narrower with depth. This tendency may be combatted by swinging the bucket against the wall as it drops, by hand trimming, or by making the top of the shaft enough oversize so minimum width will still be had at the bottom. It is helpful to equip the bucket with side cutters or corner teeth for this work.

Applications. A clamshell has the unique advantages of being able to stand on either the new grade or the old and to excavate at its own level, to a depth limited only by the length of cable its drums will carry, and at a height limited only by its boom length. It is the best of the rigs for handling bulky objects such as boulders, stumps, and logs, as chaining is usually unnecessary. It is the best tool for piling and burning loose brush and trees. It does not push or pull loose material to other positions as it digs. There are very few excavating jobs which cannot be done with a clamshell, and it is an excellent utility and odd job rig.

But it is slow. It ordinarily moves fewer yards per hour than any other rig which can do the particular job. This arises from the time consumed in closing the bucket and from the fact that the operator has minimum control over the position of the bucket, which is always directly under the boom point unless swung elsewhere. The bucket can be moved toward or away from the shovel by raising or lowering the boom, around it by swinging the shovel, outwardly by centrifugal force, and inward as a pendulum so that there is no point in its digging a range it cannot reach, but careful operation is required to get it where it belongs.

A number of different buckets are required for best handling of a variety of digging, and there is often loss of efficiency in the use of an unsuitable or compromise type of bucket.

A clamshell has the same reach as a dragline, except that the bucket cannot be cast as effectively. Since most digging is done directly under the boom point, maximum depth in "diggable" soil is determined by the amount of cable the drums will carry. Dumping height is controlled by the height of the bucket, which is variable; no diagram of working ranges is needed.

Operating Suggestions. Suggestions for clamshell operation are:

1. Keep bucket teeth sharp and built up to size.
2. Don't use more parts of line in the bucket than you need.
3. Be sure footing is solid.
4. Don't travel with a high boom—a bump may tip it back on the cab.
5. Don't swing uphill with a high boom.
6. Keep back from the edge of deep, wide cuts.
7. If machine tips dangerously, release both holding and closing lines.
8. Don't hit boom with bucket.

HANDLING A CRANE

The crane attachment may consist simply of a digging or hoist line run over a sheave on a lattice boom point, and down to a hook or clamp. More precise control may be obtained by attaching the hook to a sheave block, and reeving a line of two or more parts between it and the boom.

If the shovel does not have a live boom, the operating line not used for the hook can be reeved as a boom hoist line, and the regular boom drum left idle.

A small machine can carry a very long crane boom if the angle is kept high and the load is light.

A jib boom can be added to the regular boom to obtain greater reach and hold bulky loads away from the danger of swinging into the boom. The jib boom is hinged to the boom point and may be set straight or angled down. Its support cable is anchored on the boom.

Lifting capacity of a jib boom is usually less than half that of the crane itself. It leaves the point sheaves free so that another line for heavier lifting may be reeved in the conventional manner.

A crane designed for precision work such as setting steel usually has a foot accelerator. Enough lines are put on the hook block so that the engine can handle full loads at close to idling speed. Loads are lifted, maneuvered in tight places, and set at low throttle, and the accelerator is used to speed them up while in the clear. Still slower hoist speed may be obtained by slipping the clutch.

Loads may be lowered by slipping the hoist brake.

In cranes equipped with a worm or other slow moving boom hoist, the boom may be raised or lowered by engine power to give more precise control than is afforded by the hoist clutch and brake. This will bring the load in toward the shovel as it is raised, and cause it to move further out as it is lowered.

Fluid Clutches. Smooth operation is aided greatly by installing a fluid clutch or torque converter between the engine and deck machinery. These units allow the foot accelerator to exert complete and smooth control over raising, holding, and

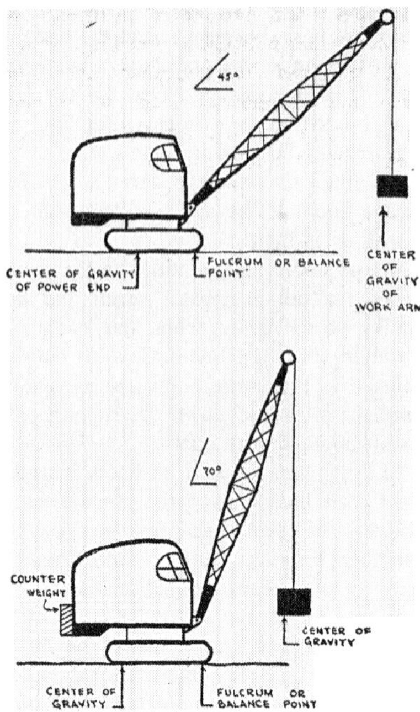

Fig. 16A. Crane balance on level

lowering. If the engine idles fluid slippage will allow the load to descend. Proper acceleration will hold it stationary and further speed-up will raise it.

If the engine is set at sufficient speed to raise the load and the hoist clutch is engaged, the load may be held by applying the brake and raised by releasing it.

Some mechanical clutches—either engine or hoist—can be slipped to obtain somewhat the same effects, but delicate adjustments are required, wear is rapid, and heat damage is likely.

Applications. Cranes lift by means of chains, slings, ropes, or tongs which grip the load and are caught in the hook. Other attachments can be used in the same manner by passing a chain around some part of the bucket or stick, or catching it in a specially made hook or bracket, and using the chain hook to pick up the load.

The crane is more efficient than other rigs for most hoisting work as it does not carry the dead weight of a bucket and other digging parts, slower and smoother lifting can be achieved by increasing the number of lines, and the operator has a better view of the hook and the load.

There are a few jobs where a digging attachment is more convenient, as the bucket can be used to move or overturn the load to make it easier to get a chain around it. Usually, however, the bucket is more of a nuisance than a help. This is particularly so when working space is limited.

Work can sometimes be done best by reeving the crane as a dragline, attaching the hook to the ends of the hoist and drag cables and leaving off the bucket. This permits in-and-out control of position without adjustment of the boom angle. However, the drag cable makes it unsafe for the ground crew to work between the load and the shovel.

Cranes can carry loads from place to place. Those mounted on rubber tires do this better than crawlers. When mobility is more important than height of lift, a shovel dozer may be preferred for this type of work, particularly in tight quarters.

Stability. The lifting ability of a crane does not depend primarily on engine power but on balance. The balance point or line is the end or edge of the tracks, tires, or outriggers toward the load. The weight of the shovel itself tends to hold it flat on the ground, and that of the load, most of the boom, and the force exerted while raising the load all tend to tip it. Weight on either side becomes more effective as its distance from the balance point increases. The entire machine functions as a first class lever with the fulcrum at the balance line. See Figure 16A.

Since the ground contact of most cranes is longer than it is wide, they have greater stability to front and rear than off the sides.

The balance of a crane can be improved by adding counterweight (but not enough to encourage straining its structure), by raising the boom so that the load is brought in closer, or by using extra long tracks or outriggers.

Putting the crane on a slope will shift its center of gravity downhill, increase the effective boom height uphill, and decrease it downhill. More weight is placed on the downhill edge so that sinking into soft ground at that end may make the machine slope greater than that of the ground. Slanted ground will therefore increase stability against uphill loads, and seriously diminish it downhill.

Crawler cranes are rated to lift 75 per cent of their tipping load in the least stable direction at their maximum boom angle (usually 75 to 80 degrees above horizontal) when standing on hard level ground. Tipping is said to occur when shovel weight no longer rests on the outer track rollers, even if the track itself is still on the ground. Draglines and clamshells are usually rated at 67 per cent of tipping load because their burden may be increased by suction or by digging beyond the boom point, and they often work on less stable footing than do cranes.

Rubber tired cranes may be rated at 85 per cent of tipping load as they have more reserve stability at the tipping point.

High Boom Hazards. A crane can lift maximum loads only if the boom is high enough to place them close in. It can handle lesser loads with greatest safety and convenience if they are close enough so that there is no question about stability. The strain on the very vital boom hoist line and compressive forces on the boom itself are greatest with a low angle and least with a high one.

It follows that it is customary to operate cranes with their booms held high. This practice involves two special dangers—the boom falling back on the shovel and overturning the shovel by abrupt swinging.

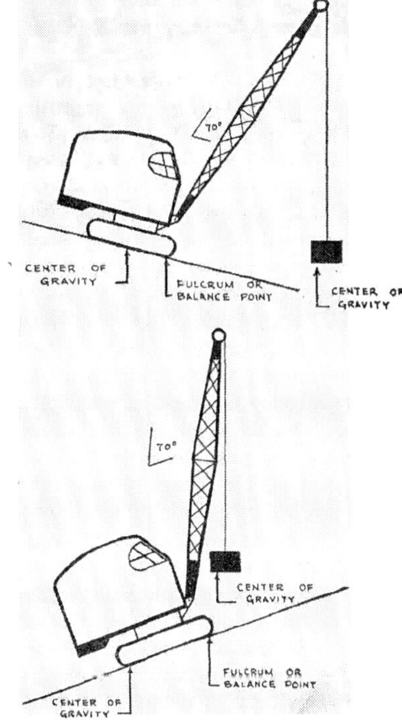

Fig. 16B. Crane balance on slope

A boom may fall back on a shovel if it is raised past its center of gravity; or as a result of swinging it from a downhill to an uphill direction, walking the machine onto an upgrade, pitching as it walks or swings on uneven ground, or recoil from a dropped or released load.

The primary responsibility in avoiding this disaster—and that is what it is—is with the operator. There are two safety devices, however, which will afford some protection against his mistakes.

The boom hoist clutch may be equipped with an automatic release which disengages it when the boom reaches its maximum safe angle. This will provide only against pulling the boom over by power when the crane is level or working downhill.

The boom may be connected to the A-frame by a telescoping stop, consisting of a pair of rods or tubes sliding inside tubes. One member is hinged to the boom, the other to the frame. When the boom is lowered this device extends automatically. When it is raised the parts telescope until they hit stops at maximum height which prevent the boom from moving higher.

This device will prevent the boom from falling back under most conditions. However, if a careless operator continues to hoist after meeting the stops he is likely to bend or collapse the boom. Also, if a very heavy load is set down while the boom is against the stops, the release of tension in the boom cable will cause it to shorten and bend the boom back against the stops.

When a low boom is swung or stopped abruptly, as is common practice in excavator operation, it tends to twist the shovel chiefly in a horizontal plane and will do no damage unless the footing is slippery or unstable. But if the boom and load are high the force approaches the vertical and tends to overturn the shovel. If it is already operating near its lifting capacity, a rough start or stop to a swing may set it on its side, or crumple the boom before the operator realizes what is happening. Increase in boom length in proportion to crane weight increases the danger.

Also, a fast swing will cause the load to move outward by centrifugal force, increasing its leverage and the danger of tipping. An attempt to stop may make matters worse. Lowering or dropping such a load while coasting on the swing is the best procedure, but circumstances may make this dangerous or impossible.

A crane carrying a maximum load should not be backed away from it, but can be moved toward it as the reaction to propelling torque shifts weight away from the direction of travel.

Danger of hitting the boom with the load increases with boom angle. A load will swing inward because of swinging and stopping forces, because of lifting clear when outside the boom point, or when it is being transported on rough ground.

Overhead Obstructions. The operator must be alert to the possibility of overhead obstructions. These are usually trees or wires. Electric lines are by far the most dangerous.

Contact may be made while working, or by wabbling into them while walking on rough ground. Unless hit by a falling wire, the operator is usually safe in his cab, even when the machine is in contact with high voltage, but he will not be comfortable. If he leaves the machine he must jump clear so as not to touch any part of it and the ground at the same time.

Avoiding Other Accidents. A crane operator usually has a much more responsible job than an excavator operator. He customarily works with a ground man or crew who will be endangered by his mistakes and he is liable to be near buildings or other valuable property. The loads he handles may be both expensive and fragile.

Most accidents are caused by falling or swinging loads. Slings or other fastenings can break or slip off. The crane operator should be fussy about both their strength and their manner of fastening.

He should inspect his cables frequently and change them as soon as they are under suspicion. He should be particularly careful about the boom cable, as a falling boom is many times more destructive than a load alone.

When a load is swung it is difficult for either the operator or ground crew to judge exactly the path that will be taken, particularly if the object is irregular in

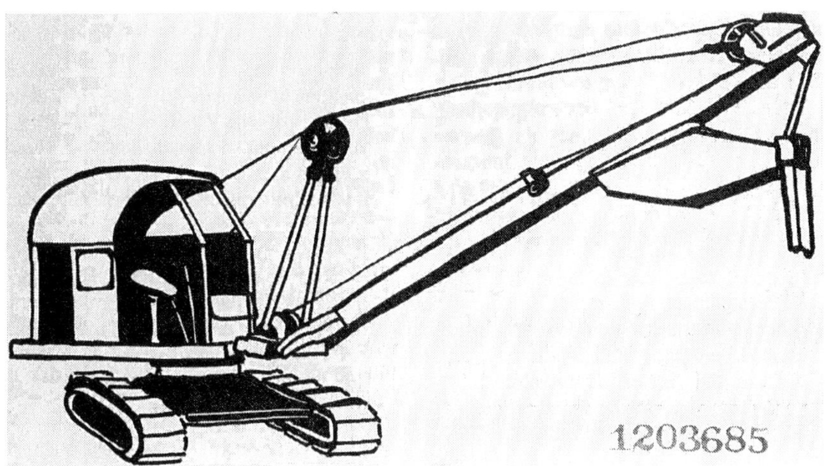

Fig. 17. Skimmer shovel

shape. Its distance in or out can be affected by centrifugal force, by soft or uneven ground under the machine, and by settling of the boom against a defective brake. Therefore, constant care is needed in close work to make sure that the load does not hit workmen, or knock objects over on them.

A variation of the falling load is one that is dropped or set down in a hurry to avoid tipping. Such action can have serious or even fatal consequences, but even so may be less hazardous than overturning the crane onto men working near it.

Questionable loads should be tested before being lifted and swung. This can be done by lifting barely off the ground with the boom in the least stable position—over the side or downhill—with the boom lowered to give a radius about 25 per cent greater than that which will be used in handling. Another procedure is to lift the load, then boom out (down) until the machine starts to tip. A safe lifting radius is about 80 per cent of tipping radius for crawler cranes, and 90 per cent for rubber tired machines.

USING A SKIMMER. The skimmer attachment, illustrated in Figure 17, is less popular than the other rigs described. It uses a boom and jack boom similar to those of the pull shovel, and these may be interchangeable, with field alteration of the boom accessories.

In digging the boom is held parallel to the new surface desired and the bucket pulled forward by the digging line. The angle of the teeth and the weight of the boom hold the bucket in the digging. When the bucket has filled, or when it reaches its extreme forward position, the digging clutch is released, its brake held, and the hoist clutch engaged so that the boom is lifted. The machine is swung to dumping position and the in-and-out placement of the bucket is regulated by releasing the digging brake enough to let it slide downward and backward along the boom, or by engaging the digging clutch to move it upward and outward. The bucket bottom is hinged in the same manner as the dump bottom pull shovel bucket, and it is dumped by a mechanical trip.

The bucket is then allowed to slide back along the boom to position for digging and stopped rather abruptly with the digging brake so that the door swings shut. This can be done during the return swing.

This rig is best used for comparatively shallow excavation and finds its greatest use in removing old street paving, horizontal coal layers which have been uncovered by stripping overburden, or any other material to be dug in layers too thin for efficient dipper stick work. Because of the horizontal bucket motion it leaves a smooth subgrade with minimum effort or skill on the part of the operator.

It digs powerfully once it can get under the layer being removed but has difficulty cutting down, so that a starting cut must sometimes be made by other machines. Because of the fixed angle between its teeth and the work it does not handle boulders successfully, and on the whole is not flexible enough for general excavation. A skillful operator on a dipper stick can duplicate skimmer performance so closely that the skimmer is a good investment only if a large amount of work within its narrow range of ability is available.

The skimmer attachment is not available for the majority of popular shovels, but is manufactured on special order in capacities from three-eighths up.

CHAPTER TWO

CONVEYOR MACHINERY

BELT CONVEYORS

The belt conveyor is a transporting, elevating, or distributing machine made up of an endless flat belt which carries a load on its upper surface. It operates between a head and a tail pulley and is supported by idlers, which are supported in turn by a frame. These conveyors are made in small portable elevating units which are loaded with hand shovels, and as giants which carry millions of tons of earth for many miles.

As independent units, they are well suited for rapid transportation of loose material. They have less mobility and flexibility than trucks and scrapers, and are therefore used chiefly where large volumes of material are to be moved along one route. They are particularly applicable where the load must be lifted steeply, or carried across rough country where road construction would be difficult. They are desirable as feeders for processing plants because they provide an even and continuous flow. They simplify traffic problems where hauling space is restricted, as in tunnels and in busy pits. However, they are not adapted to hauling big chunks which clog hoppers, damage the belt, and are likely to fall off in transit.

Their mechanical efficiency is high, as very little dead weight must be moved with the load, friction is at a minimum, and power-consuming starts and stops are rare.

Persons used to seeing soil move by the bucketful or truck load have difficulty appreciating the volume production represented by the thin ribbon on a belt.

A conveyor can be kept in touch with a receding pit face by fitting in portable frame extensions (this can be done quite easily in some of the newer mine-type shuttle constructions) and splicing in extra belt, by installing a feeder conveyor between the hopper and the face, or by trucking from the excavator to the belt.

In addition to their use as independent and semi-independent haulers, conveyor belts are used as parts of loading, ditching, and processing machines.

The large permanent type of belt conveyor is almost unique among machines used in connection with excavation in that it is usually custom built and rules of design and construction are flexible, so that it can be "tailored" to the job, and to the customer's needs or whims; and there is no clear distinction among the functions of design, operation, and maintenance.

The belt itself is the most important single item. It is an endless flat strip of rubber-covered cotton or rayon fabric laid up in plies. The type of fabric and the number of plies determine the strength of the belt. The rubber cover provides no strength, but protects the fabric from abrasion and weather. Its thickness and quality are varied to suit different types of service.

In an average installation, the belt is about half of the original construction cost,

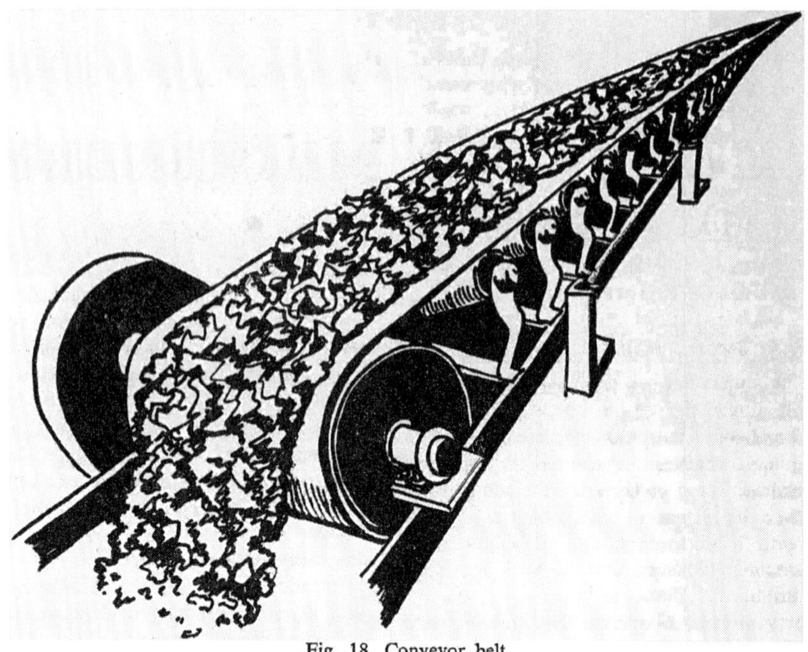

Fig. 18. Conveyor belt

and its repair and replacement is the biggest maintenance charge. Proper care of it is therefore very important.

The belt extends between a head pulley (which may be the drive pulley), and a tail or return pulley, and carries its load on its upper surface, usually toward the head pulley. Its upper strand is supported by idler sets whose three rollers are arranged to shape it into a trough, and the lower strand is supported at wider intervals by flat rollers called return idlers.

Drive. Power carries from the drive pulley to the belt by friction. If the resistance of the belt to moving is greater than the friction, the pulley will spin or slip inside the belt, with resultant loss of power, and wear on both surfaces. The amount of friction or traction is determined by the nature of the surfaces, the slack side tension on the belt, and the area of contact.

The load or tension on the carrying strand of the belt, which tends to cause slippage, is made up of the pull of gravity on this strand and its load, friction in idlers, pulleys, in the belt, and in its load; and the inertia of the whole system when starting or accelerating.

The drive pulley surface may be bare metal, or be covered by smooth face or grooved rubber lagging. Such lagging may be bolted or vulcanized in place. It increases traction particularly when the belt is wet or frosty, and prevents pulley wear.

The belt is held in full contact with the pulley by its tension on the slack or low tension side. This tension is normally regulated by some form of gravity takeup, in which a hanging weight exerts a pull on the tail pulley, or a special takeup pulley, which moves outward if the belt slackens, and inward if it tightens. If the incline is steep, the weight of the slack side may maintain sufficient tension. Very short conveyors may have threaded adjustments to move the tail pulley in or out.

The amount of drive traction which may be obtained by increasing tension is limited by sharply increased power requirement and shortened life of a too-tight belt.

The area of contact is determined by pulley diameter, the arc of contact, and the number of pulleys. A thicker pulley not only increases the contact area, but also reduces flexing strain on the belt. Its disadvantage is the cost of the pulley itself and of changes in frame and layout to accommodate it.

If an existing drive pulley is replaced by a larger one, the belt will then be driven at higher speed with less power, unless the gearing or the motor is changed. Putting on lagging has the same effect.

A belt whose strands are parallel will have 180° of contact with the drive pulley. This contact can be increased by a snubbing pulley on the slack side. Increasing the degree of wrap in this manner is the cheapest way to increase contact area.

Head pulley drives are adequate for short conveyors, and for those so steeply inclined that the weight of the return strand maintains a high tension on the slack side. For conditions requiring greater traction, tandem pulleys are used, as in Figure 21. Up to 440° of wrap can be obtained in this manner.

A belt which is pulling a load changes shape as it goes over a drive pulley. It is stretched thin where it first contacts it, and then fattens up as its tension is reduced. It

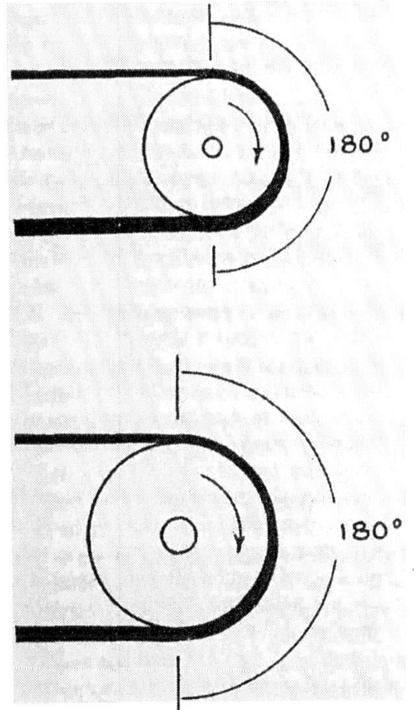

Fig. 19. Drive pulley contact

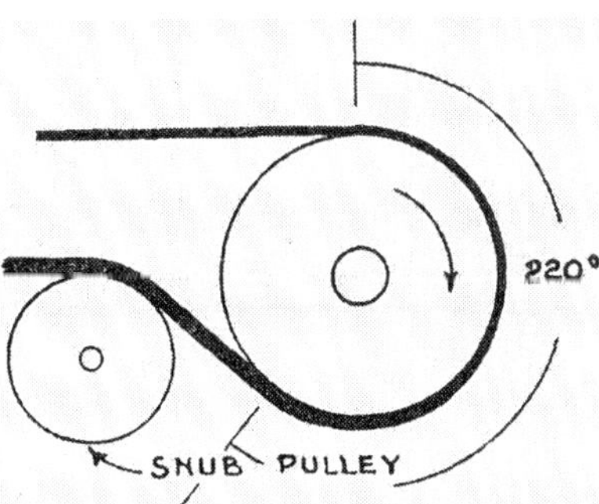

Fig. 20. Snub pulley

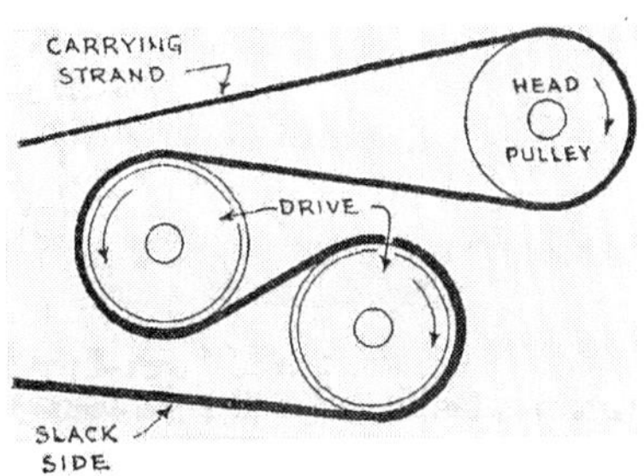

Fig. 21. Tandem drive

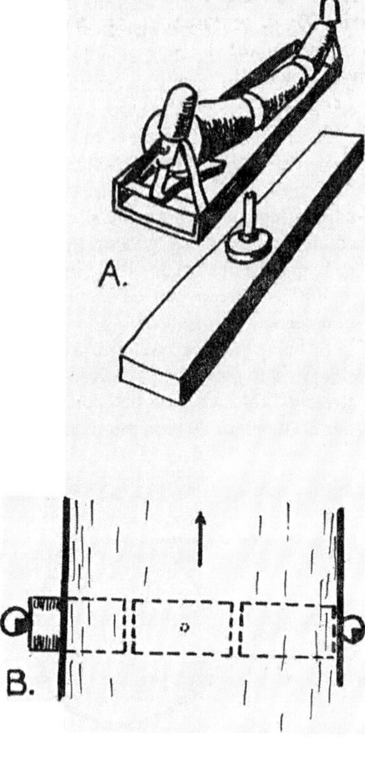

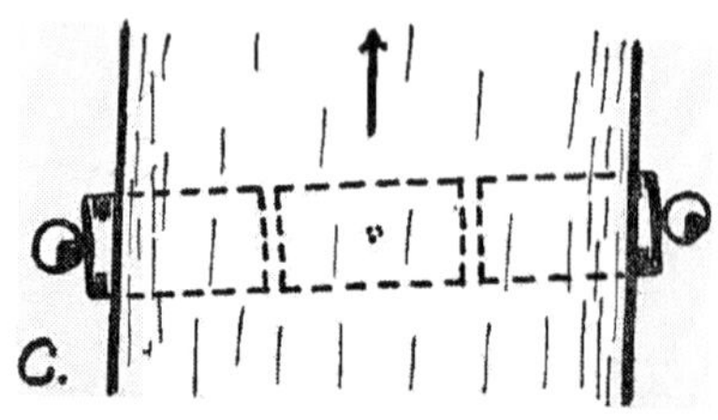

Fig. 22. Training idler

moves fastest where it is thinnest, in the same manner that water is accelerated in going through a restricted place in its channel. The change in belt thickness and speed is quite small, but requires that the second of the tandem pulleys turn a bit slower than the first, if extra stress is to be avoided. The amount of difference varies with the load. Where electric drive is used, separate motors on the two pulleys can be made to automatically adjust their speeds to each other.

Alignment. The belt is sensitive to tiny changes in frame and pulley alignment, which will cause it to wander out of a straight line. Internal changes in belt tension, or a splice which is beginning to pull apart, may have the same effect. Trouble with wandering can be greatly reduced by making both pulleys and frames wide enough so that the belt does not have to run absolutely straight to keep out of trouble. The wider construction is more expensive, of course, but through the years it should more than pay for itself in longer belt life, and in reduced checking and adjusting.

If the framework is out of line, the carrying strand may still track well enough, being steered by the load seeking the idler troughs. The return strand, however, will find the shortest path, or allow itself to be influenced by sloping idlers, so that it will rub against stationary parts. Unfortunately, it is almost standard practice to carry the return strand inside the framework, where it is very difficult to see just what it is doing. It is much better to hang it below the frame members, where misbehavior can be readily observed and necessary corrections made before serious damage is done.

Troughed idlers will steer the belt if they are tipped in the direction it is moving. However, only a very slight tilt can be used, as if it is overdone it will set up a drag against the bottom of the belt which will wear it rapidly and consume extra power.

Self-aligning or training idlers are mounted on a center swivel, and have vertical spools set at each edge. If the belt rubs against a spool, it tilts and presses a lined brake shoe against the adjoining roller, slowing it, swinging the idler, and shifting the belt back toward center. This device creates very little drag, and if placed every 50 feet in place of regular idlers, will keep a belt in line under any ordinary conditions. Two placed at 30 foot intervals ahead of the tail pulley will line the belt up properly to go under the loading point.

Loading. A belt is loaded by a chute from a hopper or another conveyor, or by a feeding device. The material always has some vertical drop, and may be moving in a different direction and at a different speed than the belt. As a result there is impact to the belt in stopping its fall and giving it its proper speed and direction. Chutes and baffles should be so constructed that they convert as much as possible of vertical and cross movement into movement in the direction of belt travel, and so that they deposit their load squarely in the center of the belt. If the material is abrasive, the chute may be so constructed that a layer of the material piles up on its slope and protects it. When the main conveyor is a big one, it may pay to put in a short feeder conveyor (transition belt) to insure feed at the proper speed and in the right direction. This transfers wear from an expensive belt to an inexpensive one.

Course lumpy materials, particularly when they include sharp edges, require greater precautions at the loading point than fine or soft ones do. Impact damage to the belt can be reduced by using thicker rubber on both the top and bottom surfaces, and by using closely spaced rubber-cushioned idlers at the loading point.

A certain amount of belt wear from impact and abrasion occurs each time the belt goes over an idler, as it is a high point with a sag on each side. The slight but steady wear at these points can be reduced by minimizing the sag, which is affected by belt tension and idler spacing. Maximum smoothing effect can be obtained from any given number of idlers by using wide spacing near the head, where tension is greatest, and progressively narrower spacing toward the tail pulley.

A fact that is commonly overlooked is that any internal movement in the load on the belt is absorbing power. Such movement cannot be stopped, but it should be kept to a minimum.

If the load slides back on itself going up an incline, a lighter load may be more efficient. If it slides on the belt, a belt with a cleated surface may be required. Often a change in the moisture content of the earth will cause or cure this type of trouble.

Spillage. There is usually very little spillage off a belt which is wide enough for its load. However, many belts are overloaded in bulk if not in weight, and the jouncing over idlers, movements in the load, and changes in alignment result in pieces rolling and bouncing off the sides. If decking is placed under the full length of the belt, or sometimes under the ends and the loading point, this material will be kept from spilling onto the return belt. A clean belt and clean pulleys and idlers mean longer belt life.

It is important that a diagonal cleaner bar or plow be placed on the return strand just above the tail pulley, as a large sharp object might otherwise fall on the belt and catch between it and pulley, so as to puncture the belt or even cut it into strips.

Wet dirt and many other materials stick to belts and will build up to considerable amounts if permitted to do so. A scraper may be placed just after the discharge point, to clean the belt and drop the scrapings into the same receptacle as the main stream.

This cleaner may consist of a rubber or steel blade, or a series of them, or be a serrated rubber roll or bristle brush which revolves oppositely to the belt. Proper pressure against the belt and adjustment for wear are provided by counterweight or springs.

A water spray, with or without squeegees, may be used alone or in conjunction with a mechanical cleaner.

Skirts are used to prevent spillage off the sides of a belt at a loading point, and to increase the capacity of short belts. They usually consist of vertical or sloped faces of metal or wood, with a lower piece of flexible rubber which makes light contact with the belt. This rubber should be lowered as it wears, and replaced whenever necessary, to avoid damage from material wedging under it.

Dual skirts have an inner board to form the load, and an outer member to prevent fines from seeping through.

Safety devices. A belt conveyor will run for long periods without attention. Idlers require lubrication twice a year or less (if of the modern type with bearings and seals), a well founded and constructed frame will keep the parts in line, and properly designed and protected chutes may never clog. However, sudden accidents happen which can be very costly if their results are not controlled, and it is impossible to keep a man always on the alert for events which may not happen for years, or ever. Controls should therefore be installed which will automatically react to emergencies.

If the power fails on an inclined belt, it will tend to run backward, jamming its load into and around the loading chute. This movement can be prevented by a holdback ratchet on the head or drive pulley, which will allow free working motion, but immediately lock against backsliding.

If objects jam between the belt and a pulley, if discharged material backs up into the belt, or if the belt runs off its rollers the power requirement will be sharply increased. If drive is electric, an automatic overload switch in the line can be set to cut off the power, and prevent or limit the resulting damage.

It is also a good plan to use a motor which is not too big for its job. 200 horsepower can drag a belt into a lot more trouble than 100 can, without increasing current requirement as sharply. The oversize motor will also put greater stress on the belt when starting it with a load.

A moderate rise in power consumption without increase in load indicates increasing friction. It may be in dry or broken idlers, result from the belt rubbing the frame, or be caused by too much tension. A record of current consumption will often show up such conditions before they would be observed otherwise.

OPERATING CABLE EXCAVATORS

Slackline. Most slacklines consist of a tower connected to a lower anchor by a heavy track cable, which can be tightened or slacked off by a winch operating through a lighter cable (the tension line) and pulleys. A dragline-type bucket is suspended from the track. It can be pulled to the tower by a two-speed winch, and tends to move toward the anchor by gravity. When pulled close to the tower it is dumped automatically into a hopper or onto a pile.

To start the digging cycle, the drag brake is released and the bucket allowed to coast out to the digging point. The fastest trip is obtained by holding the track cable at such tension that the bucket will just clear the ground or water as it reaches the digging point. This may require a tight track for distant digging or a slack one for working close to the mast.

When the bucket reaches the digging it is checked with the drag brake and the tension brake is released, allowing the track to sag and the bucket to rest on the ground. If the pit is under water the operator must get used to the feel of the controls and the angle at which the track cable enters the water when the bucket is making proper contact.

The low speed or drag clutch is engaged and the bucket pulled in. In good digging it will fill within a few bucket

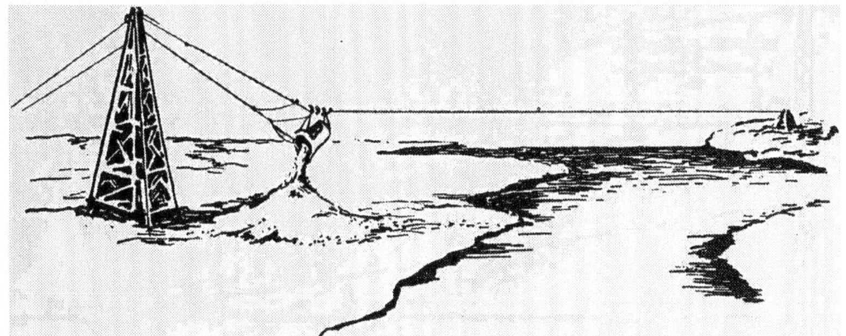

Fig. 23. Slackline

lengths. The tension clutch is then engaged so that the track tightens and lifts the bucket clear. The high speed drag or inhaul clutch is then used to pull the bucket in to the automatic dump.

The maximum load of the cycle occurs when the bucket is being hoisted and pulled in at the same time. For this reason the high speed drag is usually not connected until the hoisting is complete.

When a pit or a line of digging is opened the cut should be made at the far end and worked toward the mast. This provides a slope, as in Figure 24, which allows the bucket to operate at maximum efficiency. If excavation is selective, similar cuts and slopes can be developed in each deposit.

In caving or flowing material, digging can be continued on one line for a long period. The tail anchor can be moved a considerable distance each time it is reset.

If the material does not cave at all the bucket may cut a slot wider than itself because of wabble, or exceptionally, cut so narrow a slot that it cannot be deepened because of jamming between the sides. Tail anchor moves under such circumstances must be short, but should be sufficient to place the new line of digging on undisturbed ground in the outer area to minimize trouble from skidding sideward into the former cut. Ridges left standing between the two cuts may be taken out by moving the anchor halfway back after completing the second one.

The digging ability of these buckets is comparable to that of draglines and the ability to handle hard formations increases rapidly with size. The carrier prevents them from bouncing and losing contact as much as a dragline bucket, but the inability to vary the direction of digging makes it more difficult to cope with boulders or seams of harder material sloping across the digging line.

Bottomless Buckets. A bottomless crescent bucket may also be used with this type of slackline excavator. Such buckets are generally used for open stockpiling, but can dump in a hopper that is partly buried in the pile, or which has a hard surfaced ramp to carry the bucket to it.

The bucket is placed to dig in the same manner as the standard cableway bucket. However, having no bottom it cannot be raised from the ground without dumping its load, so it is dragged all the way in, usually with the low speed clutch.

Drag Scraper. A drag scraper generally has a mast at the low side of the pit and a tail anchor which can be moved around the high side. The bottomless bucket is dragged by an inhaul line reeved through a pulley mounted in the mast to a winch drum. It is pulled back by a winch (back-

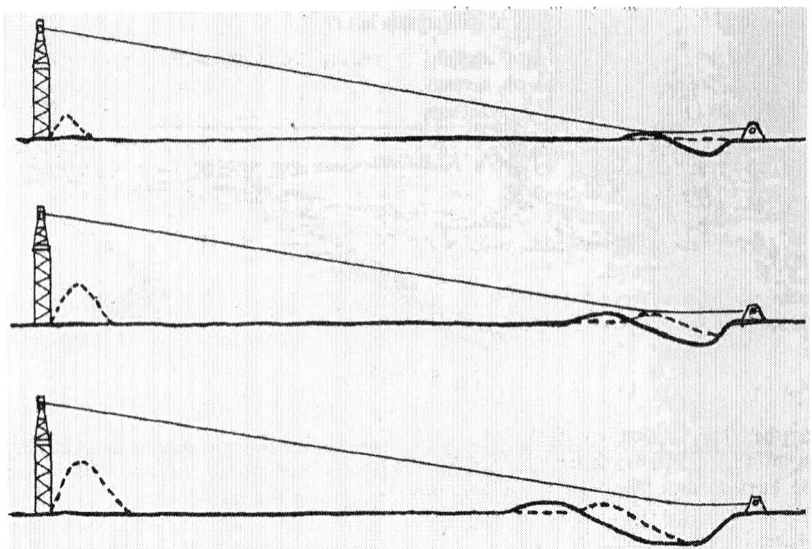

Fig. 24. Starting a slackline pit

haul) line reeved through an anchor pulley.

The bucket is moved out to digging position by engaging the backhaul clutch. This automatically releases the backhaul brake. The rear of the bucket is shaped so that it rides on the surface without cutting in.

To dig, the inhaul clutch is engaged, the bucket fills, then automatically rides up on its load so that it does little additional cutting. If it does continue to cut, and offers heavy resistance, it is probably not properly designed for the material and should be altered or replaced.

The bucket rests on the ground as it is pulled in. It is dumped by pulling it over a hopper, or by releasing the inhaul clutch and pulling it backward.

In deep soil which caves or flows into the bucket slot, the same line of digging can be used for long periods. However, it is better practice to shift occasionally so as to lower wide sections fairly evenly.

If the soil stands up in banks, frequent shifts must be made as the existence of narrow, deep slots will make working difficult. If a rapid shifting device is used at the anchor end, the whole working width is taken down in layers, usually starting at the back and taking successive strips toward the hopper.

If the tail block is shifted by hand, each move should be long enough to enable the bucket to start a new path without sideslipping into the previous one.

When peak demand for spoil is in excess of average production, digging may be done at short range for quick returns and at the back of the pit for more leisurely stockpiling.

Uses. Both slacklines and drag scrapers are at their best in digging extensive areas of uniform material. Such conditions are often found in surface mining of sand, gravel, clay, or mine tailings. Large units may be used effectively in removing blasted rock, if breakage is good, and the bucket path can be kept parallel to the face.

Material may be dumped into conveyors, sorting and processing plants, or trucks, or stockpiled. Certain types of cut

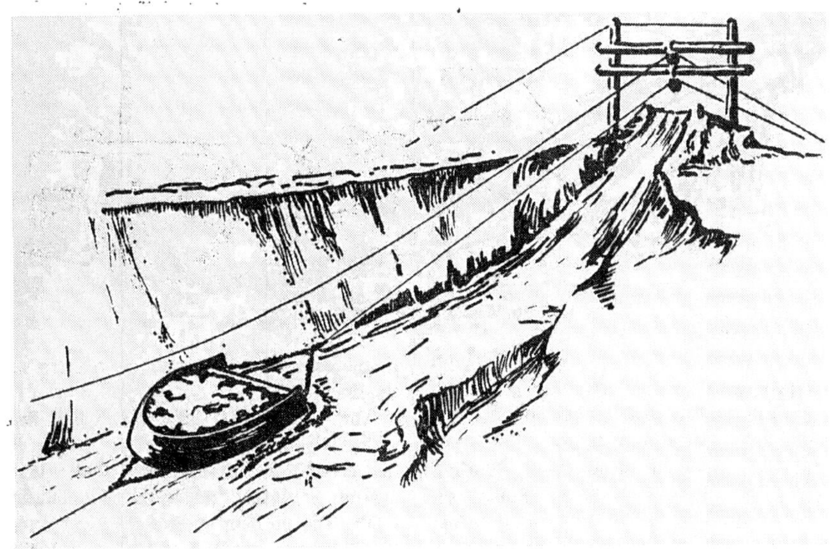

Fig. 25. Drag scraper

and fill jobs, such as building levees and embankments, can be handled by use of mobile head and tail towers.

They find an important use in rehandling stockpiles. Material brought to a yard by trucks or railroad cars can be high piled and fed into hoppers as required. Drag buckets may be reversed on the cables so that inhaul and outhaul functions are exchanged to pile material and later reclaim it.

These machines are not suited to most types of selective digging, to soil containing boulders too large to go into the bucket, and to jobs where there is not sufficient yardage to justify the cost of setting up. In general, they are not good investments unless assured of enough work to pay for them, as they may be difficult to adapt to other jobs.

In suitable work, they are highly efficient because of the small number and light weight of moving parts.

A drag scraper installation is less costly than a slackline. It is equally satisfactory in many jobs, but when the haul is long, or the digging deep under water, the higher speed of the slackline makes it more desirable.

LOADING WITH A EUCLID. The Euclid loader, Figure 26, is made up of a heavy frame carrying a plowshare which loosens soil that is picked up by a belt conveyor which discharges it high up and to the side. The conveyor is driven by a rear mounted engine.

Movement of the loader depends on a towing tractor. It acts as a semi-trailer, will turn at angles sharper than 90° with a radius of about 37 feet, and it can be backed.

The loader drives are controlled through three levers mounted on the back of the tractor seat. One lever operates the clutch which connects or disconnects the engine from the belt drive, another regulates the tilt of the edge, and the third the depth of cut and height of carrying position.

For the first cut the ground should be smooth enough for easy operation of the

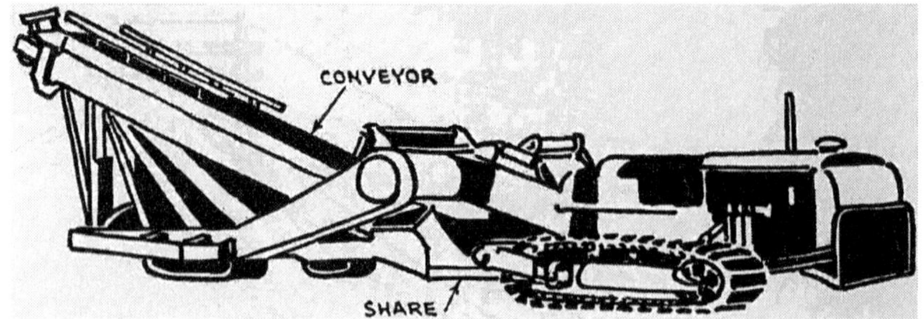

Fig. 26. Loader and towing tractor

towing tractor. If it is very hard it should be loosened with a ripper, or a pusher tractor should be used behind the leader. For best production, the working strip should be long to avoid losing time during turns, and trucks should be routed to face in the direction of digging while being loaded.

The tractor is moved forward in low gear, and the share edge is lowered until the cut is heavy enough to make the tractor engine work hard, but not enough to make the engine lug down or the tracks spin. The elevator belt is started as soon as the dirt reaches it. A truck should be under the discharge chute in such a position that the rear of the body will be loaded first. The truck is then allowed to fall back relative to the chute so that the front is loaded. Then, as the hauler moves out from under, the load is topped off, front to rear, for maximum heap.

When the truck is filled, the tractor and the belt are stopped until the next truck is in place under the chute. Forward motion and belt movement are resumed simultaneously, and stopped when the next truck is filled.

If the tractor has a torque converter forward motion can be started and stopped by opening and closing the throttle.

At the end of the cut the knife is lifted clear of the ground and the machine is turned around. A fresh cut may be started on the opposite side of the digging area, or the return may be made to overlap on the first cut.

The plowshare is nine and a half feet wide. Digging with the whole length in the ground consumes more power than making a deeper cut with part of the blade. The first cut must be full width unless the knife is tilted. After a good digging depth is reached in the first one or two passes a layer is taken off with side cuts. An average cut is four square feet, which may be two feet wide and two deep, four feet high and one wide on a side cut, or any other arrangement.

The conveyor is at the left side of the plowshare so that dirt dug on the right side must be deflected onto it by the moldboard. Digging on the left side therefore consumes less power than on the right.

Normal working speed is the tractor's full throttle in low gear. In easy digging, with a steady supply of large trucks, production may be 900 yards per hour. A greater yardage can be sidecast.

DITCHING WITH A WHEEL. Getting in. A wheel ditcher, Figure 27, works by lowering a turning wheel equipped with buckets to the ground and walking away from it. The buckets cut the earth and carry it forward to the top where they dump most of it on a conveyor which carries it to the side and piles it.

In starting a cut, as the wheel is lowered to the ground the buckets will start

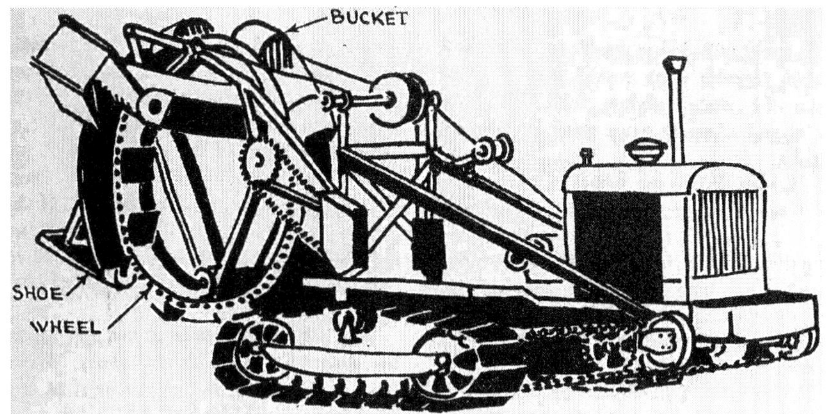

Fig. 27. Wheel ditcher

to dig. The machine is stationary. Enough weight should be allowed to rest on the buckets so that they will fill heaping without gouging deep enough to slow the wheel.

If no shoe is used, the wheel can be lowered to cut bottom grade before walking the machine. The ditch will have a rounded beginning, as shown in Figure 28 (A). Position should be such that the center of the wheel is over the starting point of the full depth ditch.

If a shoe is used, the rear hoist is pulled high and the wheel lowered in the position shown in (B). When the buckets cut so far that the shoe rests on the ground, the ditcher is moved forward. The rear cables are left slack so that the rear weight rests on the shoe and only enough tension is kept on the front cable to prevent the buckets from biting too hard.

The wheel will cut to an increasing depth as the supporting shoe is pulled into the ditch, and the front lift cable is unreeled to allow the wheel to sink.

When bottom grade or the limit of depth of cut is almost reached, the wheel is held from further down-cutting by holding the front cables, while continuing motion. The shoe will move down the ramp until it levels out. The rest of the cutting at that depth is done with rear cables slack and front ones tight, unless it is the full depth to which the machine will dig in which case both cables can be slack.

Control of the machine is complicated by the fact that the digging center is behind the tilting and turning center of the tractor. If the tractor starts up a grade or over a bump, a shoeless wheel will dig down and will have to be raised to keep level. If supported by a shoe, it will not cut down and may be raised slightly.

In starting down a grade the tractor will pitch forward. The wheel will tend to rise out of the ground, and either the front or rear cable should be slacked to drop it to grade. If a shoe is used a slight pitch might not require any attention as

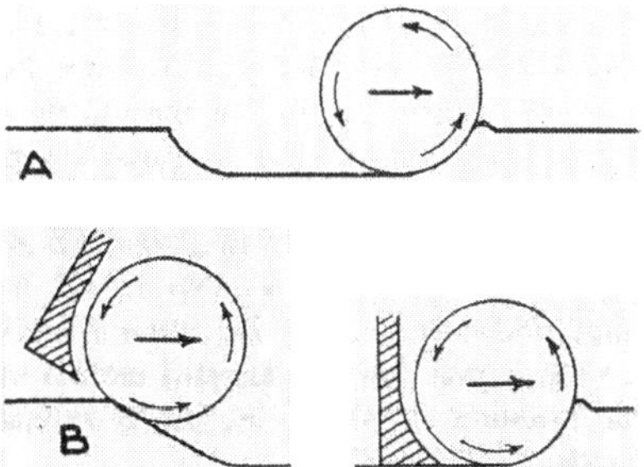

Fig. 28. Starting wheel cut

the back will lower itself on the slack cable. Greater pitch would involve dropping the wheel slightly with the front hoist and allowing extra slack in the rear cables.

Digging. When the wheel is at the correct depth, the machine is moved forward just fast enough to keep the buckets reasonably full. Crowding too hard will overwork the engine and put a heavy strain on the digging parts without adding particularly to the output. If there is a safety clutch on the digging mechanism it may slip excessively and heat under overload.

Selection of the right combination of gears for various types of digging must be based on experience. If either the machine or the soil conditions are unfamiliar, it is good policy to err on the side of underfilling the buckets rather than overcrowding.

Soft rock usually responds best to a high wheel speed with very slow walking speed. If dirt is very soft it can sometimes be crowded so that material in excess of the bucket capacity piles on each side of the ditch without damage.

Obstructions. Where boulders, heavy roots, or pipe lines are liable to be met, both walking and wheel speeds should be low and the engine should be throttled down at critical points.

Soft boulders will be cut through by the teeth. Hard ones may be pulled up to the surface. This is less likely to happen if the boulder is deeply buried, because the deeper boulder not only is held down by a greater weight of dirt, but the direction of the tooth contact tends to force it forward rather than up.

The wheel will usually ride over a deep stone it cannot move, so that it will be lifted above grade. If a big stone is near the top it may stop the forward motion of the machine, in which case power should be cut off promptly.

If a boulder is pulled to the surface, it is liable to land in a very inconvenient spot, forward of the wheel and between the tracks. It may be necessary to lift the wheel into transporting position, walk forward until clear of the rock, push it out of the way, and back up until the wheel can be lowered to the ditch bottom. If the boulder is too large for the wheel to clear, the wheel drive clutch may be released so that the wheel can turn as it is pulled over it.

It is sometimes easier to cut and repair tile lines than to work over them. Where they are expected, tile of proper sizes and joint fillers should be kept on the job.

Where a pipe must be left intact digging should be done cautiously, and the wheel lifted to inspect the digging face whenever contact is suspected.

When the wheel is lifted above grade to clear any obstruction, it may be worked back to grade at the other side in the same manner as the cut is started.

Turning. Turning must be done with great caution while digging, as the whole machine revolves around the center of one track, which causes the wheel to move sideward in the earth. If the buckets are equipped with long side teeth or side cutting bars are used, and the earth is reasonably soft, a gradual turn can be made without damage. However, too sharp a turn may bend the wheel frame or the wheel itself, or pull the wheel frame off the vertical track.

If a shoe is used, it further limits the ability to follow a curve, and if very long, such as a tile laying shoe, may prevent any significant turning unless it is loose enough, or hinged, so that it can trail instead of being swung wide.

Line and Grade. In most ditching work, it is important to both keep the machine accurately in line and working at the proper depth. Since the operator cannot see his ditch bottom, surface controls must be used.

The ditch is first surveyed and the course and the depth of cut ascertained. A line is then established at a fixed and constant distance above the bottom grade, and offset from the center beyond the track line of the ditcher.

The line should be established at a height which will put it at least a few inches above the ground at all points, stakes driven along it, and the exact height marked with nails. A string is stretched along these nails.

A rigid bar is fastened to the front of the power unit of the ditcher with one end over the string when the ditching wheel is centered on the ditch line. A plumb bob or other weight is fastened to the bar so that it will hang directly over the string.

The operator can then keep the machine on correct line by keeping the plumb bob just over the string. If the ground is irregular, the cord holding the plumb bob can be run through eyes or pulleys so that the operator can reach an end of it to raise and lower it when necessary.

The same apparatus fastened on the side beam of the wheel can be used with a fixed length of string to determine both line and depth.

Targets. Targets, shown in Figure 29 (A), can be used instead of the string. Each consists of a light pipe that can be pushed or driven into the ground, and a red and white strip on a collar which slides up and down the pipe and can be locked to it by a clamp screw.

The pipe stakes are set sufficiently far from the center line to avoid interference with the machine. The strips are set at some arbitrary distance above grade which should keep them at least four feet above the ground surface.

The device shown in (B) is attached to the wheel frame, preferably forward of the center of revolution. It consists of a bracket that will offset the device from the frame, a rigid vertical pipe, a rod made of smaller pipe or round bar inside it, with a double clamp, one part of which prevents the rod from sliding down in the pipe and the other holds it from turning.

A telescoping horizontal arm is fastened rigidly to the vertical rod. Sliding and turning adjustments are made so that the end of the horizontal arm is the same distance from the center line as is the target line, and the same height above the bottom of the lowest wheel bucket as the target strips are above the ditch line.

The operator can sight the row of targets and his indicator from his position on the side of the machine. If the indicator is not in line the machine should be steered or the wheel raised or lowered until it is. When a target is reached the swivel clamp

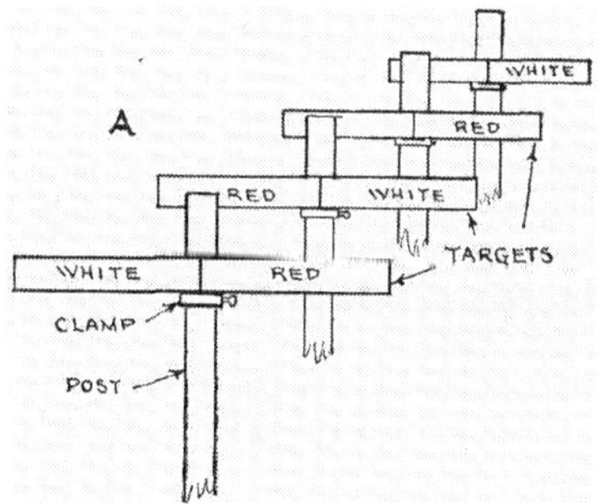

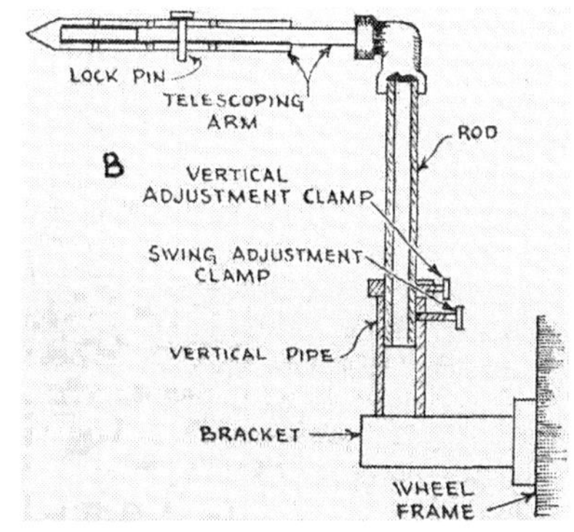

Fig. 29. Targets and sight arm

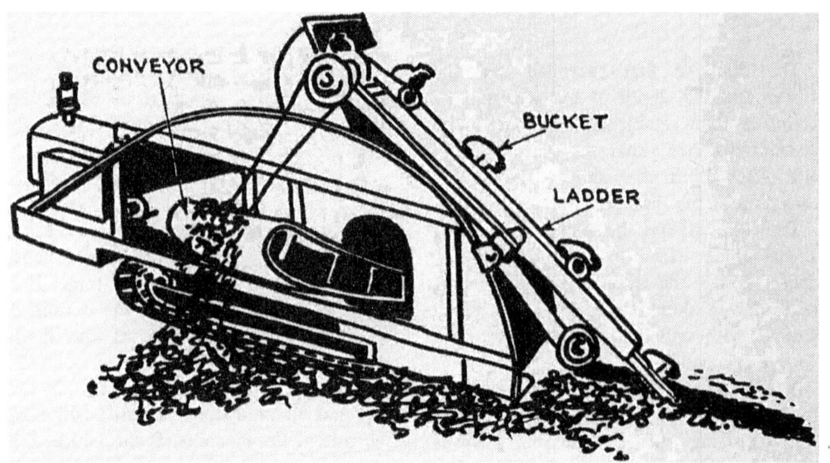

Fig. 30. Ladder ditcher

is loosened and the indicator swung back so as not to hit it.

A slight tilt of the wheel forward or back will not affect the accuracy of this gauge noticeably, but if the wheel is tilted considerably the indicator will be moved downward so that it will have to be held under the target line for accuracy.

Targets are best suited to work in straight lines on constant gradients. If changes in direction or grade occur, a target must be located at each point of change.

DITCHING WITH A LADDER. A ladder ditcher, Figure 30, carries a chain of digging buckets on a boom. In general, operation is similar to that described for wheel ditchers.

A ditch can be started with an almost vertical rear face by holding the machine with the steering brakes, engaging the bucket and conveyor drives, and lowering the boom. When full depth is reached, the machine is driven forward at such speed that the buckets will be kept full without straining or gouging.

Digging can also be done in reverse, in which case the dirt is cut off the rear wall and carried under the tail pulley. The buckets will not load as well and a considerable amount of spillage will be left on the trench floor. The elevator must be centered so that the tracks can straddle the ditch. It may be necessary to put planks across the trench, between the elevator and the tracks, to prevent caving.

Reverse is generally used only in undercutting pipes, curb stones, or other obstructions. The ditch is dug forward until the obstacle is reached. The elevator is raised to clear, the ditch dug a little past it, the elevator lowered, and cutting done in reverse right up to it. The slanting slice of earth under a pipe which cannot be reached may fall into reach of the buckets or be knocked down by hand.

HYDRAULIC DREDGING. Hydraulic dredges consist of floating pumps which suck in mixtures of soil and water and deliver them to a disposal point through a pipe line. They can excavate underwater soil or waterside banks, transport the soil mixed with water through pipes, and rough grade it on fills. A single machine with proper accessories may perform all three functions, or it may dig without transporting, or transport from a hog box without digging.

A light cutter dredge with the deck house cut away is shown in Figure 31.

Advancing. The dredge is advanced into the digging by the pull of the swing winches and the pivot action of the spuds. One of these, called the working spud, is down while the dredge digs; the other is used for advancing. The swing lines go around sheaves near the end of the suction ladder to anchors placed at each side.

In Figure 32 (A), the dredge is headed directly toward the bank with the working spud down. Pull on the left (port) swing line, as in (B), moves the cutter in an arc to the left, pivoting the dredge on the working spud. Pull to the right (starboard), (C), pivots the dredge to the right.

These movements are made with the pump operating and the cutter head revolving. Firm material is taken in slices from the top down by lowering the ladder at the end of the swing. Flowing or sliding soil may be dug at the toe only.

When the material within reach of the cutter has been dug to bottom grade, the ladder is lifted until the cutter is above the top of the bank and the dredge swung to starboard, in the (C) position. The walking spud is dropped, the working spud raised, and the dredge pulled to the left, as in (D). This causes the dredge to pivot on the walking spud, moving the working spud forward a distance determined by the extent of the side swing in (C) and (D).

The working spud is then dropped, the walking spud raised, and digging resumed.

Spuds must penetrate the bottom sufficiently for a firm grip. If the bottom is hard extra heavy spuds may be required, and it may be necessary to raise and drop one of them several times in one place in order to sink it.

Anchors must be raised and moved forward occasionally to avoid back pull from the swing lines.

Cutting. The standard dredge does its hard digging while swinging toward starboard, with the cutter blades slicing up into the bank. The swing back to port is usually made with the cutter at the same level, cleaning up loosened material.

The revolving of the cutter causes it to act as a driving wheel, pushing the ladder to the left. This push will be weak in loose sand and strong in hard formations. On the right swing the winch must overcome this resistance in addition to crowding the blades into the digging.

When swinging to the left, the cutter may pull the ladder too rapidly for effec-

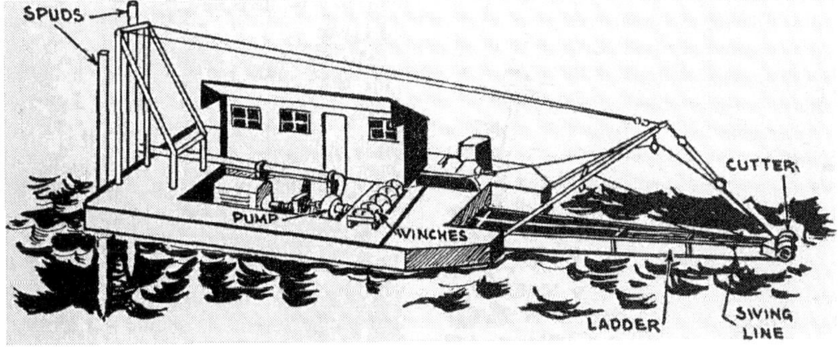

Fig. 31. Hydraulic dredge

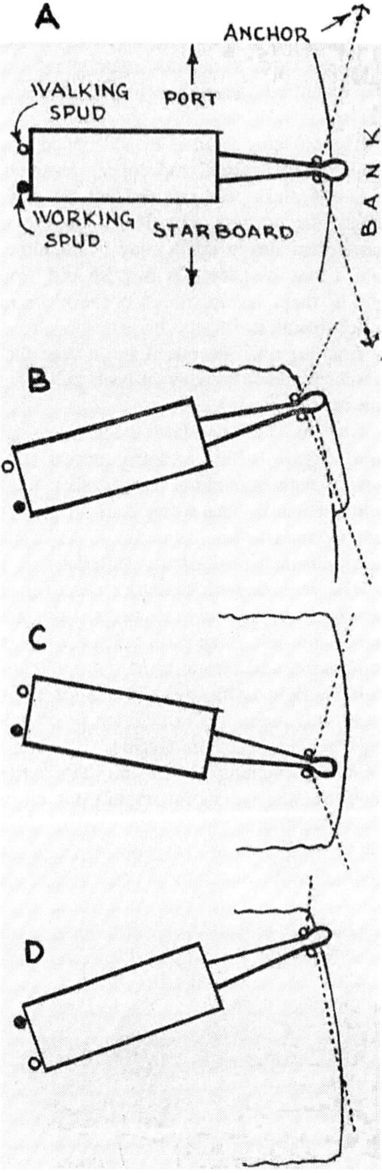

Fig. 32. "Walking" a dredge

tive cleaning and cause the winch line to run slack, and foul on the cutter. In this case the right hand winch brake is applied enough to hold the cutter back to line speed.

Dry banks are undermined by cutting below the water level so that they slide or cave into the water where they can be picked up by the suction. The height that can be safely reduced this way depends on how readily it slides and the size of the unit. If too much material comes down at one time, it may bury the ladder or damage the dredge.

It is usual for a dredge to cut to bottom grade at each stand, regardless of the depth of material to be moved. Moving the dredge with the attendant work of shifting anchors and extending discharge lines is laborious.

The percentage of solids moved is regulated by the size of the slice taken with each pass and the speed with which it is cut. The slice can be enlarged by advancing the dredge further toward the bank or by lowering the ladder. Cutting speed depends primarily on the hardness of the material and is controlled by the speed of the swing line, which can be reduced from maximum by throttling the engine, shifting to a lower gear, or slipping or intermittently disengaging the clutch. If the winch cannot pull at full speed production may be increased by turning the cutter faster.

The dirt cut or stirred up by the blades falls inside the cutter head or is carried around the outside. Water pulled in from the top and the left side by suction of the pump further breaks up the pieces and carries them up the suction pump.

A skillful operator will keep the intake of solids as high as their nature and the velocity of his discharge line permit without plugging. But he should not allow the line to plug as cleaning it out is liable to be a long and tedious job involving down time that will far outweigh the gain of extra few percent of solids that caused it. A blocked line is particularly serious in freezing weather.

A vacuum indicator connected to the suction line indicates roughly the percentage of solids since these are heavier than water and a higher vacuum is needed to lift them. A plugged intake will cause vacuum to rise sharply. This reading may also be affected by the angle of the ladder, being somewhat lower when taking off the top of a high bank than when working on the toe.

A pressure gauge attached to the discharge pipe indicates the head against which the pump is working. This will include the lift from the pump to the discharge opening, and the friction against the pipe walls, which is ordinarily a function of the length of the pipe and the velocity of the stream. If the line threatens to plug discharge pressure will rise abruptly as the stream backs up behind the obstruction.

A threatened plug may be cleared out by alternately speeding and slowing the pump, or raising or backing off the cutter so that clean water or a leaner mixture will be pumped. When working with heavy loads it may be good practice to run the pump at somewhat less than full speed so that extra force will be available to break through obstructions.

If the pipe does clog, the plug may be located by observing the small leaks usually found in shore pipe connections, or by tapping the tops of pipes with a light hammer. They will have different tones when full of water under pressure, or mud, and when partly empty.

Whenever a part of the pipe is to be cut off from the dredge to be taken apart, or when the dredge is to be shut down for more than a few minutes for any reason, plain water should be pumped until the discharge line is clean. Mud in a pipe will make it difficult or impossible to handle, and may dry out and cake in an unused line so that water will not move it again.

If a single discharge opening is used the dredge pump must be stopped whenever a new section is to be added.

The dredge operator is usually unable to see much of the discharge area, but he must fit his actions to its needs. Some quick signaling system must be used for efficient operation. Flags, telephone, and radio are suitable for different conditions.

A dredge is generally run on a two or three shift basis, because of its high original cost, and desire to spread its overhead over as many work hours as possible. It will often be run through lunch periods by a split crew.

Clearing. Vegetation should be cleared away from the path of a dredge as thoroughly as possible. Cattails, reeds, and other swamp growth may wind around the cutter head, causing constant delays in removing them. Stumps and logs may jam or break the cutter, or block the pump or line.

When the cut is across dry land, brush, trees, and stumps should be completely removed and heavy sods plowed and disked. If it is in tree or stump filled water, it should be preceded by a snag boat, generally a grapple dredge. This will raise the stumps and logs, blasting them when necessary, and swing them outside of the cutting line or load them in barges for removal.

Discharge Pipe Patterns. Dredge discharge lines can be rather readily extended at either end but are difficult to move. Floating pipe is unwieldy to manage in long towed sections, particularly in a wind, and it is a slow job to disassemble and relocate it piece by piece. Shore pipe must always be taken apart for moving. It should be flushed out with clean water first to lighten it.

As the dredge advances, the discharge

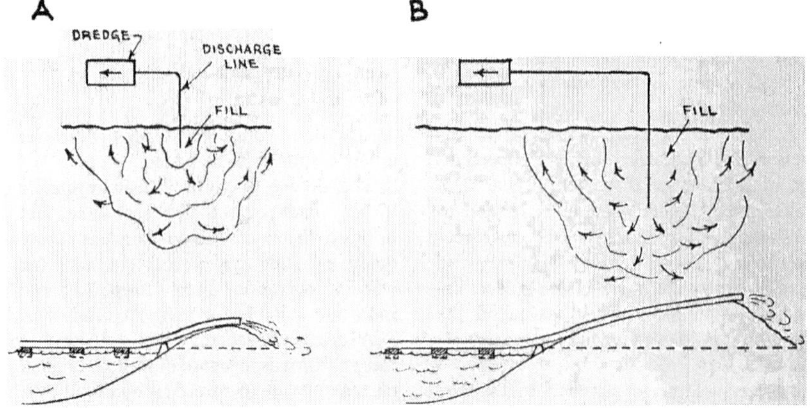

Fig. 33. Open hydraulic fill

pipe must be lengthened. This is done by disconnecting at a joint immediately behind the hull, placing another section by means of a light floating crane, and refastening.

In view of the difficulty of moving and the desirability of keeping the number of curves to a minimum, patterns involving a few right angle turns and considerable extra footage of pipe are in general use. These are indicated diagrammatically in Figures 33 and 34.

Wherever possible, two or more discharge openings with control by diversion valves are used. This system permits adding pipe without stopping the dredge.

For simplicity, a balanced amount of cut and fill is assumed in these diagrams, so that the fill advances about the same distance as the dredge moves. This favorable condition will be found when the bottom is used merely as a borrow pit so that depth or width of cut can be varied to keep pace with the fill, and when the disposal area is large or deep enough to easily accommodate the spoil from specified channel work.

If a natural balance does not exist the depth of the fill may be varied to provide for disposal of greater or less yardage. This is a matter that should be carefully calculated in advance, so that seawalls or berms (earth dikes) may be built to correct height. Maximum depth of a single layer of hydraulic fill is limited by the pressure exerted on the dikes.

If the spoil is coarse or mixed, additional layers may be added by using the technique to be described for building dams. The extra height of the upper layers may put additional load on the pump, which would have to be speeded up to maintain full output. However, the lift may be more than balanced by the use of lines shorter than those needed to spread the same material over a larger area.

Handling Pipe. The method of handling discharge pipe depends largely on the firmness of the fill surface. On quick draining sand or gravel a pile of pipes may be pushed up to the discharge end by a bulldozer, which can also place each pipe in a position to be engaged by hand. Shovel dozers, or cranes or cherry pickers mounted on crawlers or half tracks, can also be used for handling.

On softer fills, pipe may be dragged in place by winch lines or handled by a dragline on platforms. More often, men in hip boots roll pipes along the discharge line to its end and place them by hand. If footing is very bad it may be necessary to lay boards on each side of the line to support the men. Supporting boards may also be placed under the pipes.

Fills are likely to be somewhat firmer in the immediate vicinity of the discharge than further away, because coarser particles tend to settle out first.

Pipe joints are often rather out of line when connection is first made. They are straightened out two or more lengths back from the discharge opening, usually by prying into better position and blocking up. A straight line reduces friction on the load and wear on the pipe.

Leaks at joints are quite common. They can sometimes be stopped by tightening the connection or prying the pipes into better position. They can be checked by forcing in the thin ends of shingles. Those on the bottom block themselves with spoil without requiring attention unless the pipe is on a trestle.

In general, pipes lying on the ground do not need to be hooked together as the friction in the joint and with the ground holds them together. However, on curves and at changes in grade fastening is necessary.

Individual joints are fastened by winding wire around the pipe hooks, which are placed in line with each other, and twisting it tight with a short stick which is left caught against the pipe. On curves only the outside hooks need fastening.

If the line is on a trestle each joint must be fastened on both sides.

A number of joints may be fastened at a time with a cable. It is looped and clamped, or fastened in any manner to both ends of the section to be strengthened, put under the hooks on one side of the first joint, over the top of the pipe to be caught, under the opposite side on the next, and on alternate sides to the other anchor. The cable becomes tighter with each joint, and if the correct slack has been allowed, can be slipped over the last few

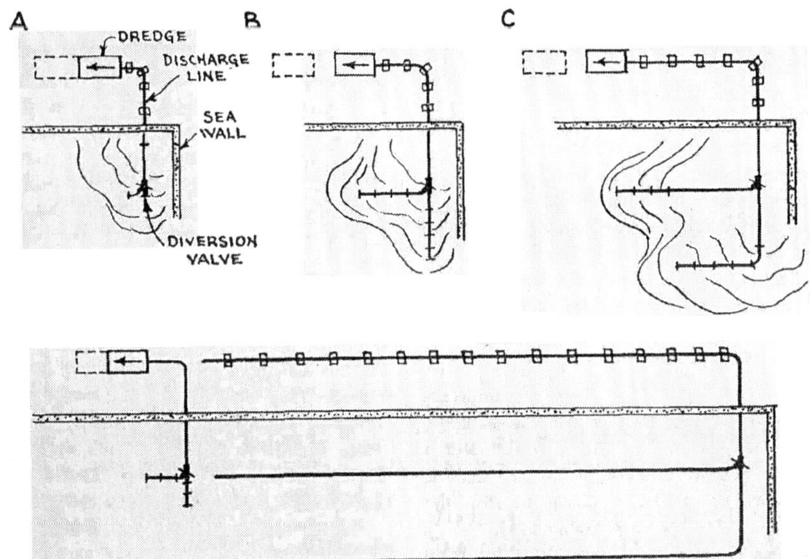

Fig. 34. Discharge pipe pattern

joints only by means of a bar. Another method is to leave it loose enough to go over all hooks easily and then use a load binder to tighten it.

Light cable, three eighths or half inch, is most satisfactory, but heavy, worn out swing lines are often used.

HYDRAULIC FILLS. Hydraulic fills may be wasting operations to dispose of unwanted spoil from channel dredging, they may be a secondary but valuable part of the work, or the dredging may be done merely to get the fill.

Open Fill. The simplest type of open hydraulic fill is illustrated in Figure 33. Spoil is being wasted on an area large enough so that the water deposits practically all its burden when allowed to flow freely away from the pipe. A small percentage of the soil is fines which remain in suspension and drain back into the dredging area.

The end of the discharge pipe is supported on a low, light horse, or a log, or any object that will keep it above the fill. It may or may not have baffles to spread the stream. The stream will first erode slightly where it strikes the ground, but deposit the heaviest particles immediately around it and lighter ones farther away. In this manner a cone will be built up. Average slopes may be as steep as 15 percent for chunky material or as flat as 3 percent for silt. A high content of solids will give a steeper slope than a light burden of the same material. The action of the flowing water will sort and pack in the pieces.

When the fill has built up to grade or to a point where it threatens to interfere with the stream, another horse is set up, a pipe length is brought forward, the stream is diverted to another line, or a signal is sent to the dredge operator to stop the pump, and the pipes are connected.

The surface may be a slope up from the start of the fill, or level. In either case it should be made as smooth as possible to simplify keeping pipe gradient and to save future grading work. If the material is so coarse that it tends to build up a series of cones, the pipe may be set on a trestle and window pipes used. If the openings can be closed a series of them can be used with open ports only in the last one or where needed. If the ports are simply slots the window pipe must be taken off and replaced at the end of the line when it is lengthened.

Settling Basins. If the proportion of suspended particles in the runoff water is large enough to constitute a serious waste or to cause undesirable silting where it is discharged, or if the edge of a fill is to be steeper than the natural slope the water must be retained or ponded in the fill area long enough to permit settlement. This is done by construction of dikes across the natural direction of flow. These may be permanent seawalls or berms thrown up by a dragline or dozer.

Berms must be of ample height and heavy enough not to be pushed out by pressure, even when softened by soaking. Dirt must be thoroughly temped or puddled around the spillway.

Spillways are usually of wood.

The amount of settlement obtained is determined by the turbulence in the settling pool, the length of time the water remains in it, and the fineness of the particles. The first two factors are largely fixed by the proportion between the area and depth of the pool, and the rate of inflow. As the fill progresses the area will be reduced. This may be counteracted for a while by raising the water level by putting more boards in the outlet gate, but ultimately additional land will have to be inundated if settlement is to be obtained.

Where the bottom being dredged includes known areas of fine and coarse

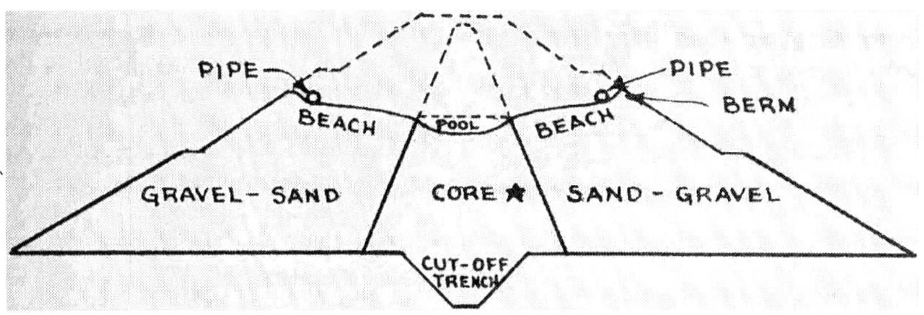

★ TYPICAL CORE CONTENTS

CLAY	20% - 30%
SILT	40% - 75%
SAND	5% - 40%

Fig. 35. Hydraulic fill dam

soils the fines may be worked first while a large settling area is available, and coarse deposits used to complete the fill.

Dams. The velocity of the soil-water mixture discharged from a dredge pipe drops rapidly after it hits the ground because it spreads over a widening area and the water soaks into the fill. The stream may immediately lose its ability to carry cobbles, drop the gravel a few feet farther on, then the sand. Varying quantities of silt and clay may be carried after the water comes to rest in a settling basin.

This results in a gradation of material, from coarse at the pipe through finer sizes with increasing distance. The separation is not clean, as some fines are trapped near the pipe and some coarse particles are rolled or carried by occasional currents well into the edges of the cone.

An earth dam must have the impenetrability of clay and the slump resistance of gravel. One way to meet these requirements is the construction shown in Figure 35. A water resistant core of clay, silt, and fine sand is supported and protected by gravel sides.

This cross section can be obtained by the proper selection and placing of hydraulic fill. The bank material must have a higher proportion of fines than needed in the dam structure to allow for the amounts left in the coarser material and that carried away in the waste water.

The dam is built up in a series of low fills. Berms are built by dragline or dozer along the outer edges of the fill and discharge lines run inside them. The spoil builds up in a slope or beach from the outer line down to a central settling basin called the core pool. Coarse particles remaining in the stream are ordinarily deposited immediately upon entering this calm water. Occasionally, concentrated runs of water may build sand or gravel tongues out into the pool. This possibility is guarded against by anchoring beams or other floats just offshore where they will break up such formations.

The core pool water is drained into a pipe line or flume which is usually on one of the outer slopes, and is built up as the dam level rises. The width of the core is regulated by that of the pool.

HYDRAULICKING. Hydraulicking is a dry land dredging operation. Banks are washed down by high pressure water jets and the mixed spoil and water moved by gravity, pumps, or both, to a fill. It is used in building hydraulic fill dams when sufficient material cannot be obtained by dredges, for breaking down banks, cleaning rock for blasting, and for various special conditions.

Plant. Hydraulicking equipment consists of water pumps, suction and discharge lines, monitors, and a flume or sluiceway to carry the mixed water and soil.

The centrifugal water pumps may be high pressure jet types, or medium pressure set in series with one pump's discharge supplying the intake of another. Usual pressure is between 80 and 150 pounds, but 275 may be used.

Steel or iron pipe is used except near the monitors, where flexible hose is required.

Monitors, also called giants, are large, high pressure nozzles mounted in swivels on heavy skid frames. Small sizes may be aimed by hand, large ones have a movable baffle which utilizes the power of the stream of water to turn the nozzle.

The sluiceway may be a gully of controlled width in the pit floor, a flume of wood or metal, or dredge discharge pipe. A gully should have a slope of at least 4 percent, but flumes can be somewhat flatter. A narrow deep waterway will give maximum velocity and transport the greatest bulk of solids.

The high pressure water may do the transporting or additional water may be supplied by low pressure pumps or gravity.

The sluiceway runs into a distributing pipe which takes it to the fill. This pipe should start high enough on the slope to give sufficient head to push the spoil through it.

Hog Boxes. If the grade is not sufficient to carry the spoil from the pit to the fill, it is poured into a hog box, which is a concrete trough from which a dredge-type pump moves it through a pipe line to a fill.

When gravity feed is not possible and the hydraulic fill method is required on the dam, dirt may be dumped through grizzlies into the hog box by trucks, washed into a sump by monitors, then pumped to the fill.

Operating a Monitor. The primary requirement of dam hydraulicking is the correct selection of material. Definite proportions of fines, sand, and gravel are usually required, and these are often not found in a single bank.

Several monitors may have to work in different sections, or in different pits, and increase or decrease their part of the output on the direction of inspectors.

If the bank will cave if undermined, the monitor should be as close to it as safety permits and the stream should be directed at the base. If most of the cutting can be done at an angle, efficiency is increased and danger from slides is reduced.

If the bank will not cave it is best cut by standing at such a distance that the stream will just start to bend downward before hitting it.

As the bank recedes the monitor must be moved forward. The flexible hose should be long enough to be laid in loops so that considerable movement is required before an extension is needed.

Sometimes draglines, bulldozers, or other machines are used to break up the soil and push it into the sluiceway.

Water jets will erode very hard soils, but not efficiently. Light blasting of clay will greatly increase production.

CHAPTER THREE

TRACTORS AND DOZERS

DRIVING CRAWLER TRACTORS. A standard crawler tractor, with the controls shown in Figure 36, may be operated as follows: Release the engine clutch by moving it forward, shift into the gear to be used, set the throttle for the desired engine speed, pull back on the clutch lever until the machine starts to move or the engine starts to work, continue the pull until the machine is moving and the clutch carrying the full engine power without slipping, then pull it back rapidly until it locks over center. Starting with heavy loads, or in high gears, requires a longer slow pull than loads easily moved. The pull should be slow enough to avoid jerking or jumping the tractor, but any hesitation beyond this minimum will increase wear.

Some engines will operate efficiently only if the throttle is wide open or nearly so. Others will perform reasonably well at any speed. On the whole, fast engine and tractor speeds are suited to rough, heavy work and fast traveling on smooth ground, and slow to moderate speeds for precise work, moving in close quarters and walking on rough ground.

The top speed of most crawlers is from four to seven miles an hour on a level, but this is in a high gear which has little reserve power for climbing hills or pulling loads. It is expensive to operate at top speed, as wear on the track chains is greater per mile of travel at high speeds, but the time saved or work performed will often more than justify the cost.

Gear Selection. For precise work, the lowest gear is desirable, not only because the slow pace gives more time to steer and make adjustments, but because the bulldozer blade or other unit will move faster in proportion to the speed of the tractor. This is because the blade speed is in proportion to the speed of the engine, while the tractor speed is varied by the gear ratio. The blade can therefore cut a grade more accurately in low gear than in second, even if the tractor speed is the same.

For less exacting work the gears should be the highest that can be used without lugging the engine below its governed speed, and which will give the amount of control needed.

Gear shifting is slow and cumbersome in most tractors because of the spur gearing used, the hand throttle setting which makes double clutching impractical, and the fact that friction in the tracks slows it so rapidly that considerable operator skill is required to complete a shift before the machine stops. The gear used throughout is therefore usually low enough to start the load, although a higher gear might move it once it is underway.

Some of the larger machines use constant mesh transmissions which shift more easily when standing, and which can be shifted on the move—an important ad-

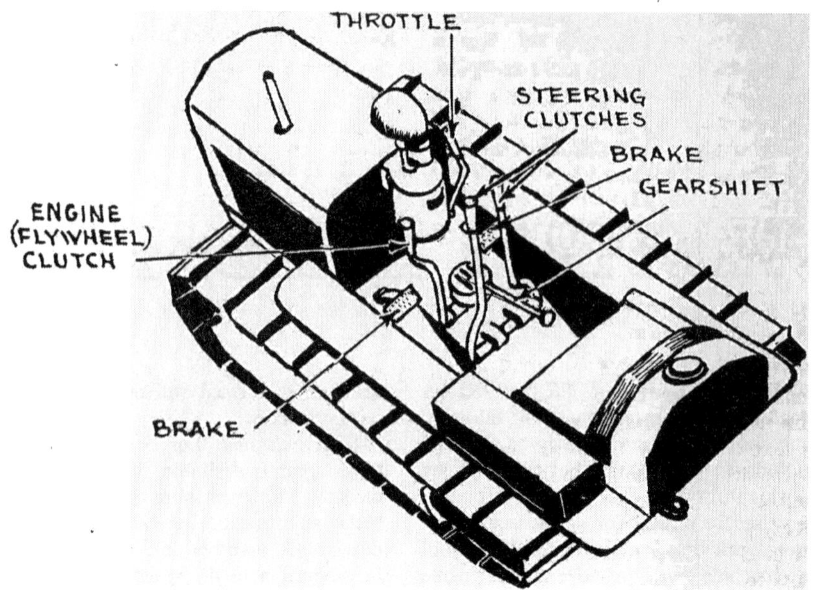

Fig. 36. Crawler tractor

vantage in scraper and other high speed hauling.

Speed may be increased or decreased by opening or closing the throttle, or by choice of gears.

Steering. Five factors are involved in steering a conventional crawler tractor—the effect of the steering clutches, of the steering brakes, of the footing, of the grade, and of the load being pulled or pushed.

If one steering clutch lever is pulled back while the tractor is moving forward, the track on that side will run idle and the other will continue to drive. The amount of turn in an unloaded tractor will depend on the resistance of the idle track to moving forward. Its resistance is made up of internal friction, which by itself will produce only very gradual change in directions and ground drag. As a result, on a smooth hard surface the machine will turn very gradually. In soft ground it will turn more sharply. If the non-driving track collides with a stump it will swing abruptly. See Figure 37.

Rolling resistance in the free track can be increased to any desired degree by applying its steering brake. If the track is locked, and traction is good, the machine will spin around it. Brake application can produce any degree of turn between this and the non-braked gradual change in direction. Braked turns are seldom smooth, as direction is usually changed in a series of steps.

When moving uphill, the free track is held back by gravity, and the weight of the machine also increases the tendency to turn, as shown in (B). Other conditions being equal, less braking will therefore be required for steering when going uphill than on the level.

If the ground is soft, both tracks will dig in. The driving track will tend to spin and dig; the braked track to push earth at the front in the direction of turn, and oppositely at the rear. At best turning consumes some power, and under soft conditions the resistance built up by the displaced earth may be great enough to prevent the turn and will put severe strain on the tracks. Under such conditions turns are best made in a series of short jerks, to give the tracks a chance to work away from the dirt ridges before turning further.

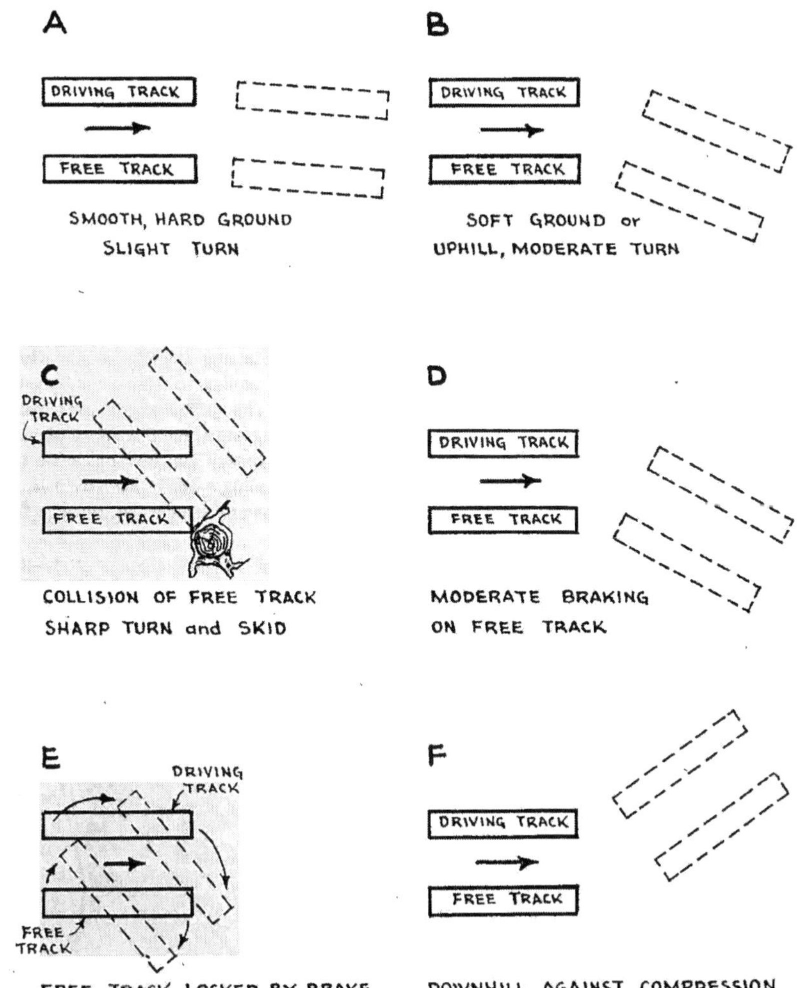

Fig. 37. Steering with clutch and brake

Since the machine drives from one track while being steered, only half of its normal traction is available. This complicates maneuvering in soft ground and with a load.

A heavy load, either on the drawbar or in front of a blade, has somewhat the effect of an upgrade in supplying a turning force when drive is from one side only; the load itself resists being turned. In general, a loaded tractor will steer more readily than a free one in response to a steering clutch only, and with greater difficulty in response to clutch and brake. Very sharp turns are usually difficult or impossible but the location and character of the load influence this.

Even with all clutches engaged, a tractor with a heavy offcenter load will tend to turn toward the loaded side.

A tractor so loaded that it is using nearly its full traction in straight going will have difficulty in turning, as the driving track will be inclined to spin. It may be necessary to relieve the load by backing in order to change direction. This is usually fairly easy with a pushed load but difficult with a pulled one.

Steering by means of clutches alone works backwards in going down a hill steep enough to make the tractor try to roll faster than the engine drives it. The effect of releasing a steering clutch in this case is to allow its track to roll free, and move faster than the other track, which is braked by the engine, so that the machine will turn away from the released clutch. Firm application of that brake will cause the machine to turn toward it in the normal manner. If precise steering is required, the brake should be applied before the clutch is released.

Steering in reverse is the same process as steering forward.

Use of either steering brake while its steering clutch is engaged will slow, stop, or hold the machine, but has no steering effect, as while the clutches are engaged the axle shaft acts as a solid unit from track to track.

There are crawler tractors which differ from the type described in important respects.

DRIVING A WHEEL TRACTOR. Operating a wheel tractor may be said to be intermediate between that of a truck and a crawler tractor. The machine is faster and quieter than a crawler of the same power, but will not turn as sharply. It can pull powerfully on firm footing, and with light loads can go forward through loose soil and mud. When moving heavy loads, however, the wheels spin readily and it gets stuck easily.

Controls are shown in Figure 38.

Steering. Steering is done by a wheel and linkage similar to that found in an automobile. The front wheels turn very sharply, giving the tractor a short turning radius. On slippery surfaces or when the machine is pulling a heavy load the wheels may skid forward instead of turning the machine.

If the foot brakes are on the rear wheels or axles and may be separately controlled, one may be applied on the side toward

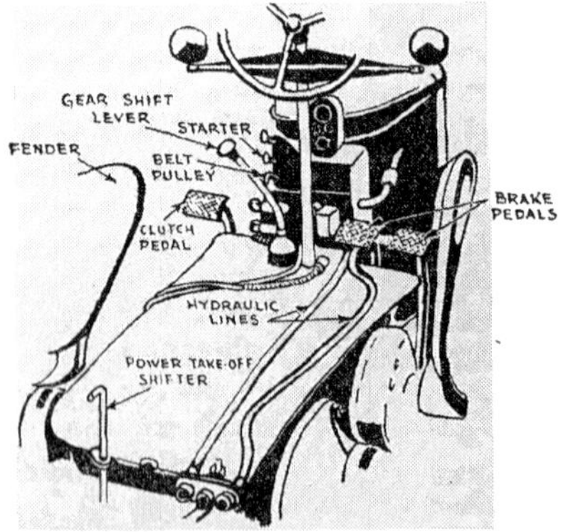

Fig. 38. Wheel tractor controls

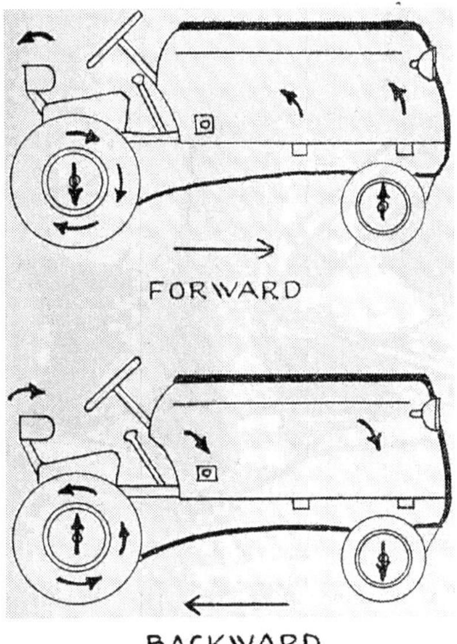

Fig. 39. Driving torque and traction

which the turn is being made. It will act as a pivot to turn the machine with the angled front wheels merely assisting. Unfortunately, many tractors lack this convenience and must be limited to much lighter loads on turns than on straightaways.

The steering wheel must be held firmly on rough ground as a bump may cause it to spin with sufficient force to break a wrist. A power booster is optional equipment on many large models, and is almost essential when a front loader is carried.

Loss of Traction in Reverse. Most rear wheel drive vehicles have better traction moving forward than backward. This is largely due to reaction from turning the drive axle, as the whole machine tends to rotate in an opposite direction from the axle.

Referring to Figure 39, it will be seen that if the tractor is driven forward, the front of the drive wheels turn downward, and the equal force of reaction from the force applied tends to lift the front wheels up and back. The weight is thus concentrated on the drive wheels.

When the tractor is moved backward, the reaction from turning the axle forces the front wheels down, increasing their load, decreasing that on the drive wheels, and reducing traction.

If the tractor is moved up a hill, this shift in weight will be affected by a shift in the center of gravity. If the hill is very steep, there is danger that a tractor climbing it forward will overturn when the power needed to lift the front wheels becomes less than that needed to turn the rear wheels.

These effects will be more noticeable when the center of gravity is near the axle, or when the tire is large so that its contact with the ground is distant from the axle. Since wheel tractors normally have short wheel bases and big tires, it follows that reduction in traction while backing will be more severe than in cars or trucks.

Speed. Top speeds of wheel tractors range from eight to twenty-five miles per hour, with those less than fifteen being the most common.

Comparison with Crawlers. Two wheel drive tractors are cheaper to buy and to operate than crawlers, get around faster, and do less damage to surfaces over which they move, but they do not have the lugging power of tracks, nor the ability to work on soft ground. They are not suitable power units for excavating hard material unless it is blasted or scarified, or for work in loose or sandy soil. They are at their best in pits, yards, streets, and other places where they can work on paved or well compacted ground; or in scraper work and other jobs where speed is more important than traction.

Four wheel drive tractors have performance intermediate between two wheel drive and crawler machines.

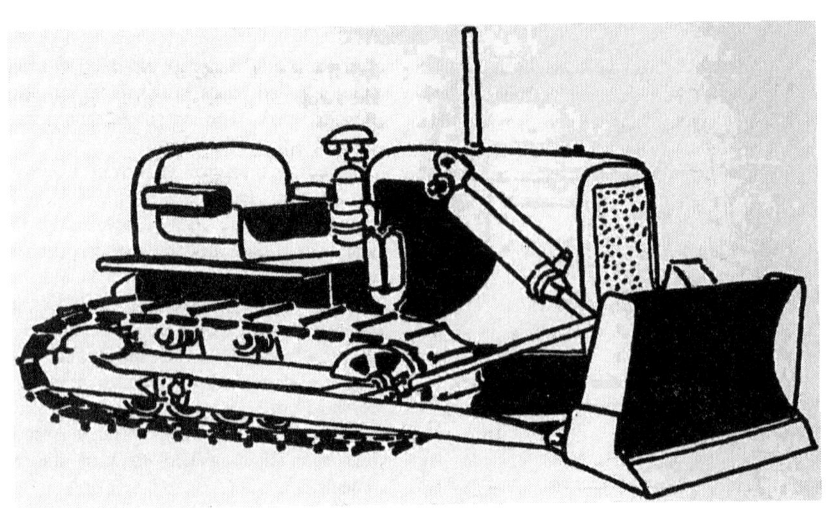

Fig. 40. Bulldozer

BULLDOZING. Digging. A bulldozer, Figure 40, is worked by moving the tractor forward, or less commonly, backward, and raising and lowering the blade to contact material to cut, spread, or transport it.

As a dozer moves forward and digs, some of the soil cut by the blade will pile up in front of the blade and move with it, and some of it will drift off the sides, forming ridges or windrows. Resistance to the machine's movement is made up of the power absorbed in cutting and breaking up soil, and in friction in the loosened dirt. If the blade is lowered, more work will be done and resistance will increase, as a thick slice requires more digging power than a thin one, and the total amount of dirt resisting the blade is increased. If the blade is raised, the slice will thin or disappear, and the amount of earth being pushed will decrease, so both work and resistance are reduced.

In heavy digging, efficient operation involves keeping the dozer pushing the most dirt it can without losing speed through slowing the engine or spinning the tracks. The operator starts cutting a slice which should give this result, and if the machine slows, he raises the blade slightly; if it is not working to capacity, he lowers it. The upward blade movement should be made as gradually as possible to avoid leaving a bump in the path of the tracks.

If the blade is set to cut an even depth, digging resistance will remain about the same through the pass, but that of the loosened and transported dirt will increase steadily. This increasing resistance does not slow the machine at first, as it causes the governor to open the throttle to maintain tractor speed. Once the engine is wide open, further increase will slow it, so that the blade should be raised gradually to the surface of the ground where it can push the loose dirt without digging more. Sometimes the blade is "pumped" during this lift by being dropped and raised quickly, cutting out a bit of extra dirt each time it is lowered.

The dozer digs and transports much more effectively downhill than on a level or uphill, and work should be arranged to work down a grade when possible.

Breaking Piles. A pile of dirt may be knocked down by walking into it with the

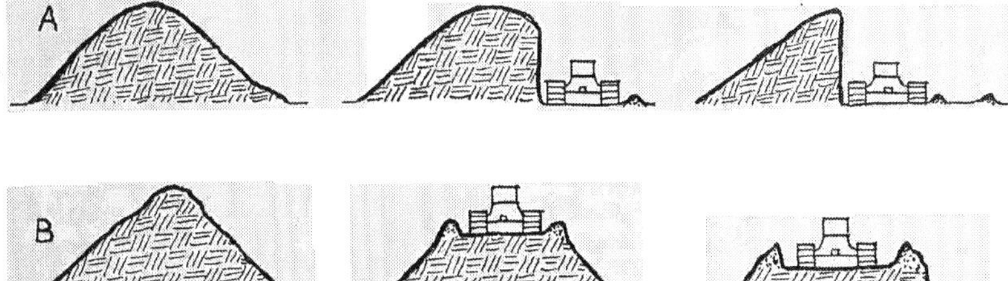

Fig. 41. Taking down a pile

blade at the desired grade, after which it may be spread or piled elsewhere. If the heap is too large or hard for the machine to take at one pass, or if it is to be spread in more than one direction, the first pass may be made to cut away part of the pile to grade, as in Figure 41 (A), or to cut the top of it part way down, as in (B). If the second method is to be used but the dozer cannot move the part it can reach, a ramp may be made by loosening the soil by pushing then backblading with down pressure, as Figure 42, so that the blade can contact the heap at a higher level.

If the pile is very large or hard in proportion to the power of the dozer, the side cut should be repeated from different angles, as in Figure 43 (A), in order to shorten the cut required for each pass. If the digging leaves a high face that might fall on the machine, the dozer should be turned toward it occasionally and driven into it with the blade held high. This should cause the bank to fall or slide without burying the side of the dozer.

If the sides are not accessible, a center cut may be made by first ramping up and cutting a slot down to grade, widening it to both sides. This slot, and any cut more than a few inches in depth, should be made somewhat wider than the blade to avoid jamming the dozer between the walls. This may happen in very narrow cuts through rocks or roots in the sides being turned by grazing contact with the blade so that they project, or by creeping

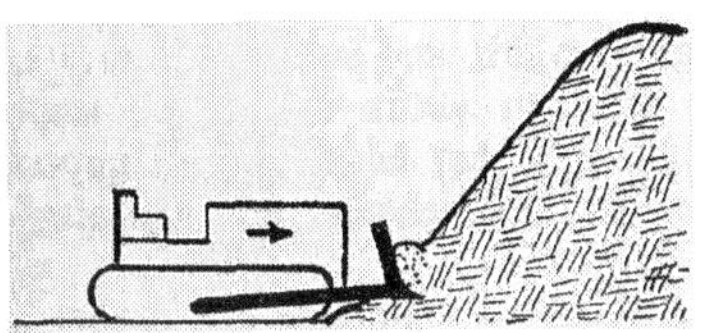

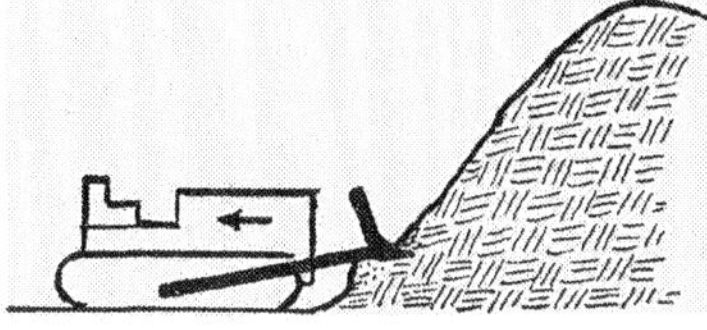

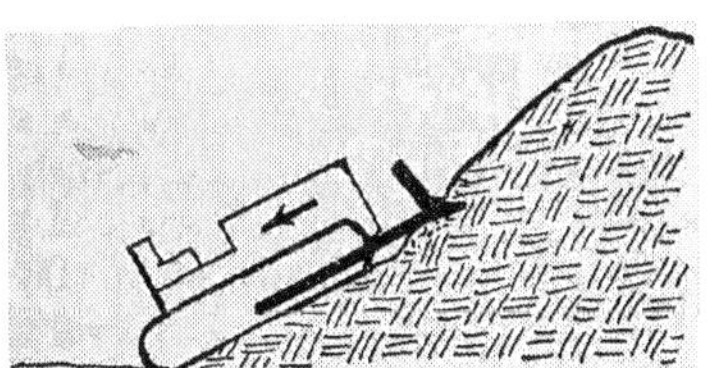

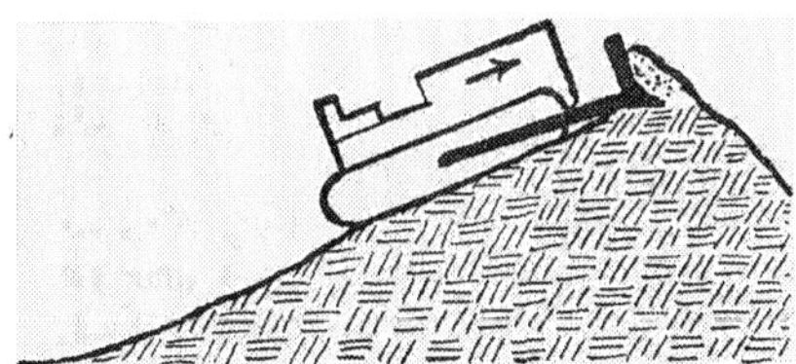

Fig. 42. Starting a slot

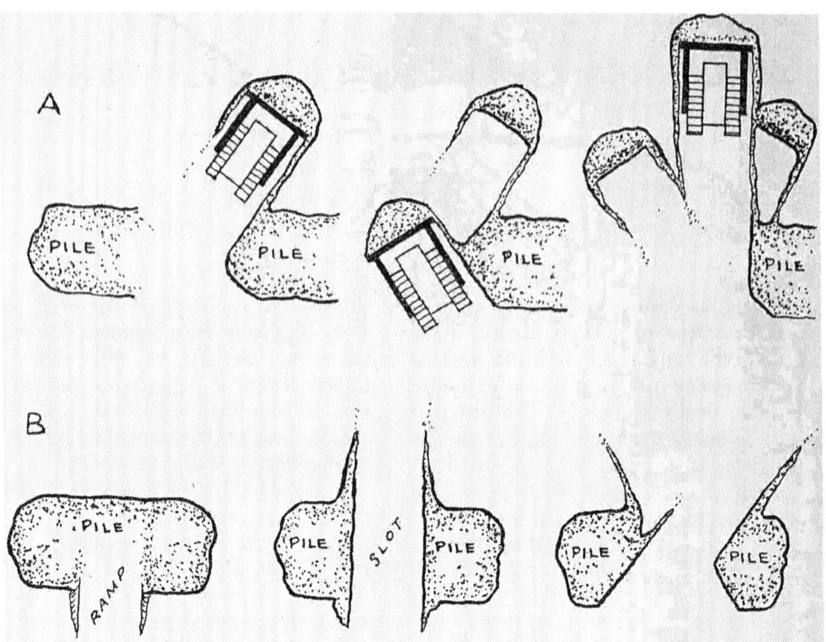

Fig. 43. Spreading pile from center

or falling in of the sides. Dozers whose blades are track width or very little wider are particularly subject to getting jammed in this manner.

A narrow cut also does not leave room to maneuver to get at a rock or other obstacle encountered in the floor.

Suck. Plastic (rubbery slit or clay) soils will pull the blade down as it is pushed through them, dragging down the front of the tractor at the same time. An adjustable blade may be set for less penetration, and overdigging may sometimes be avoided by making a number of very thin slices. More often digging is done in the regular way and gouges made below grade are refilled with loose material.

Dozers which have the cutting edge set at a pronounced forward angle and set well below the blade for a maximum suck; those with the hoist on the central frame of the tractor, and those with spring cushions on the lift rods are particularly cranky in such soil.

Road Cuts. A cut which is to have steep sloped sides, as for a highway, should be started full width, and necessary measurements made to assure cutting the slope correctly, as it may be difficult or impossible to get the machinery up on it afterward.

Side Slopes. A bulldozer not having a tilting control for the blade will cut deeper on the side which is downhill or in softer soil. On a slope, this tendency may be overcome in several ways. A shelf may be built by pushing downhill, as in Figure 44 (A)-(C), so that the dozer can start its side cut level or tipped oppositely to the slope. Or the dozer may cut and turn downhill, raising the blade, as in (D) to (F), thus cutting a more or less level shelf which can then be enlarged or graded off as in (G) and (H).

Shallow stripping may be done by starting at the top of the slope so that in each pass except the first, the upper track can be walked in the hollow made on the previous pass. This works best in pushing two ways from a central point. Stripping downhill is more efficient but it is not always possible.

Figure 45 shows a series of steps by which a short wide shelf may be cut in a hillside, working from the side only. The material is piled in the background.

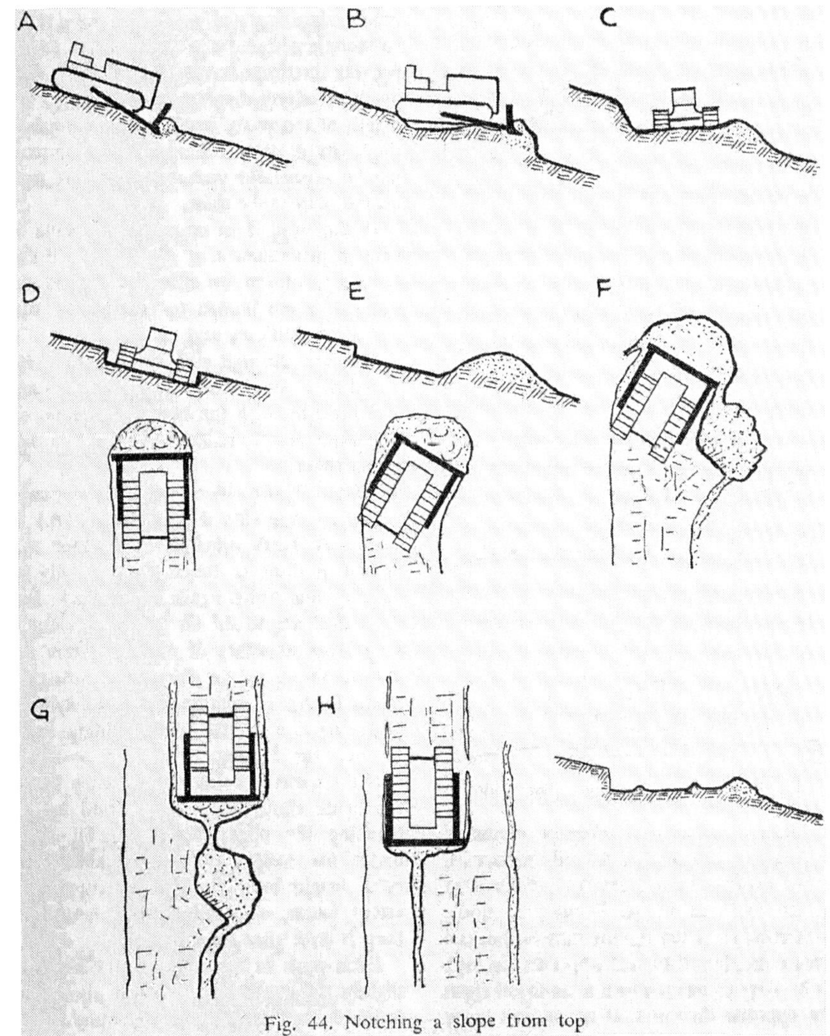

Fig. 44. Notching a slope from top

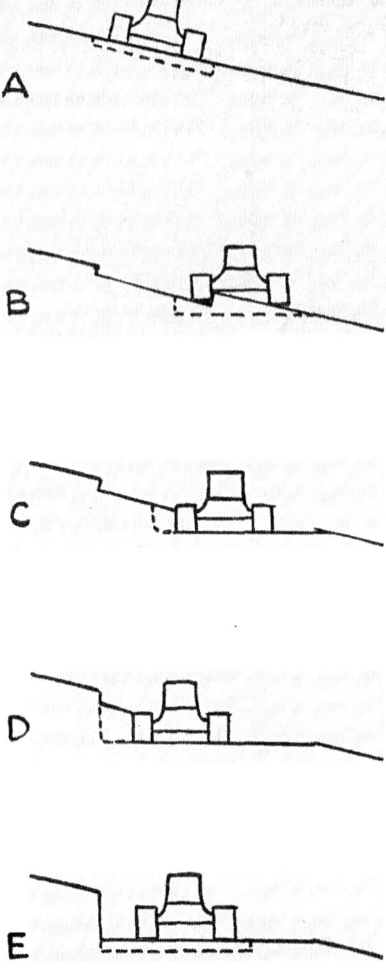

Fig. 45. Notching a slope from side

Uneven cutting may often be corrected by taking advantage of soil windrows, rocks, or other high spots to tilt the tractor up on the side where it cuts too deep. In narrow cuts, the tractor may be backed up on one side of the cut opposite the high or hard spots, or may turn so as to cut from the opposite direction. If no natural helps are available, the tractor may be walked onto boards or other lifts placed by hand at the low side, to start the cut at the desired slope.

Transporting. While transporting material the dirt which flows off the sides of the blade must be checked or replaced to keep a full load. It may best be checked by moving repeatedly in the same path, so that the ridges built up in earlier passes prevent dirt from leaving the blade. It may be replaced by digging down through the length of the push, just enough to replace the wastage, or by centering on a windrow left by a previous pass, which will replenish losses from the sides.

If the ground is smooth, the blade is held to just touch it or dig slightly. If the ground is uneven, an effort is made to cut enough in the humps to replace the dirt lost in the hollows and at the sides.

Wastage through side spill may be reduced by having two or more dozers work side by side, with the blades touching, so that little or no material can be lost between them.

Piling. Figure 46 shows two common ways of piling dirt with a dozer. (A) is somewhat more efficient as it does not involve pushing so much dirt up, only to have it slide down again at the back, but the difference would be important only if very large amounts of material were involved. Often a pile started according to method (A) is continued in the style of (B), because of the pile building back too near the digging.

If the earth is being pushed off a bank, the blade should be slowly lifted before reaching the edge. The loose fill sinks under the weight of the machine. The extra height built at the edge supplies a safety factor, and should be enough to keep it level after compaction.

Each push to the edge should be on a slightly different path, if space allows, in order to distribute its packing action.

Spreading. When spreading material, the blade is held somewhat above the original surface so that dirt can slip under it in a smooth layer on which the tractor can walk. A thin layer may be spread to the desired grade, but a thick layer should be built higher to allow for compaction. If there is not enough dirt ahead of the blade to reach the end of the area to be covered, it saves time to stop pushing as soon as the load is light and go back for more. The next bladeful will be pushed through the spot and can easily take the remnants of the first load with it.

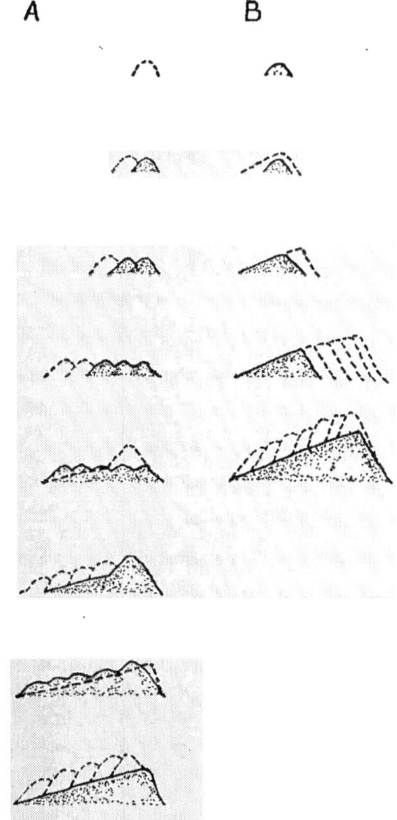

Fig. 46. Piling with a dozer

It is best to vary the path used in spreading, as it is easier to keep track of the grade if no heavy windrows are built up.

Turning. A bulldozer has difficulty turning while pushing a heavy load. More power is needed to swing the load than to push it straight ahead. As shown in Figure 47, the action is that of a bell crank lever pivoting on the braked track. Turns are easier with a wide gauge machine, as the power arm is lengthened, and harder with an angledozer or other dozer with blade carried a distance ahead of the tractor, as that lengthens the work arm, reducing the leverage.

A clutch steering machine loses half its potential traction on a turn, and a differential steering job moves into a higher gear ratio, so that, in addition to an increase in load, traction or power is usually cut down. It is therefore necessary to lift the blade somewhat on a turn and to let part of a full load escape, to be picked up on another pass.

It is often easier to break the curve up into two or more straight lines separated by angles; to pile the soil at the angle points, and to make a separate process of pushing it along the next line. The heaps at the angle points are best moved frequently instead of being allowed to pile up until difficult to dig.

Scalloping. Accurate grading is difficult on rough ground, or in rocky soil, as any pitching of the tractor is exaggerated in the movement of the blade. The blade control is not fast enough in most machines to be kept level while the tractor oscillates. If it is allowed to dig in on a drop, the material is apt to be left in a pile just beyond, and the tractor, in walking into the hole and up on the heap, will pitch even more sharply so that a series of scallops are made. Once this process starts, only expert operators can level out without going back to the beginning and doing it over.

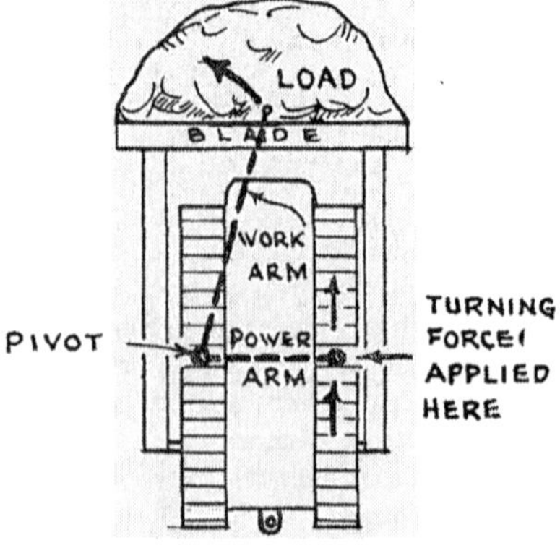

Fig. 47. Leverage in swinging a load

This scalloping is the bane of beginners and is liable to be troublesome until a feeling for the balance of the machine is acquired. An experienced operator can tell—usually without being conscious that he does so—when the machine is about to either rear up or pitch down, and start moving the blade to compensate. If the change promises to be very abrupt, he can slow the tractor by cutting the throttle or slipping the clutch and using brakes, if necessary, to get more time to change the height of the blade.

A difficulty with hydraulic dozers, and a few cable jobs, is that the drop below the line of the tracks is limited and usually is not enough to enable it to work over a peak and start a grade down the other side; or to make a sharp down cut from a level grade, except by preparing a heap or ramp to back up on, to tilt the dozer down.

Backblading. After an area has been graded it may look a bit rough because of small windrows of loose dirt, grouser prints, and piles left where the dozer turned. These may be smoothed down by backing over the area with the blade floating. It acts as a drag, smoothing off humps and filling hollows, but does not move enough dirt to change the grade. This is called "Okie dozing."

Soil that has been pushed into places where the dozer cannot get behind it can be dragged in the same manner. Hard soil or a large quantity of loose dirt is better pulled with down pressure.

Rocks—in Cuts. Rocks of flat or irregular shape may catch under the blade knife and be pushed along with it, increasing the resistance or floating it out of the ground. The blade may sometimes be freed by shaking it up and down, but it is often necessary to stop and back in order to get behind and under it.

Firmly imbedded rocks even of rather small sizes may roll or slide the blade up so that a bump is left. It may be necessary to back up, then move forward forcing the blade into the ground deeply enough to get a grip on the stone that will roll or push it out. It may then be pushed ahead a few feet, with the blade high enough to let dirt slide under it, and that dirt backbladed into the gouge.

In digging out large rocks and stumps the blade action usually is a combination of push and lift. The push is faster and more powerful but objects requiring chiefly upward motion may be handled by slipping the flywheel clutch, or releasing one steering clutch in low gear, so the blade is pushed against the object without jamming it, and lifting at the same time. This technique should be used sparingly because of excessive wear on the clutch.

Rocks—in Fills. Coarse material, such as rocks, lumps of sod, and other debris, made grading difficult or impossible. If a high fill is being made or a hole is dug to bury the trash it can be worked over the front of the fill and buried. As the blade is raised in spreading dirt the coarse material has a tendency to stay ahead of it to the last, although some will slip under

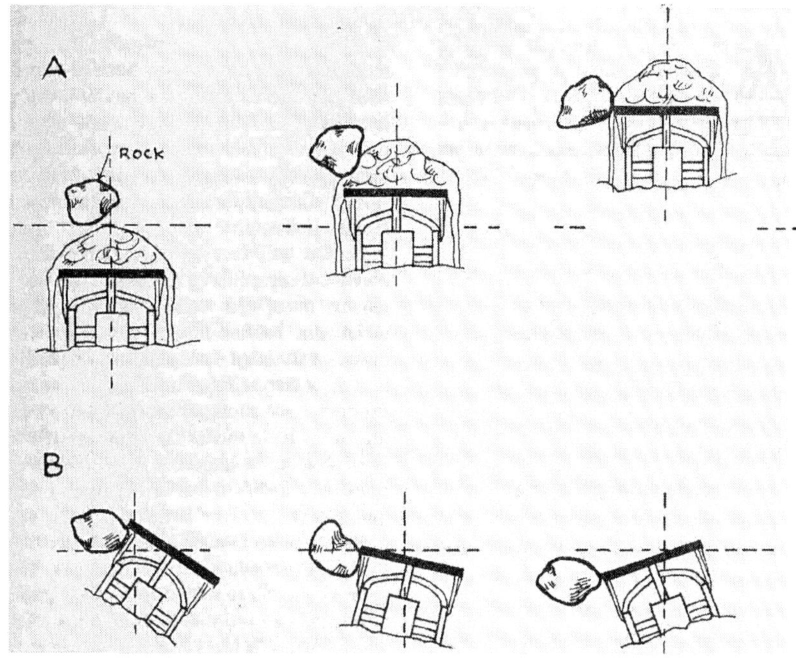

Fig. 48. Side-shifting a boulder

and sometimes force the blade up so that the grade is lost. Such pieces can be moved along a bit further with the next pass, and eventually gotten over the edge, although it is often quicker to get off the tractor and move small, difficult pieces by hand.

If the stones are not to be buried but are to be left at the side for other disposal, they can often be worked over without making special passes for the purpose. Referring to Figure 48 (A), it will be seen that if a dozer pushes a bladeful of dirt into a loose rock, the rock will tend to drift off to the nearest side. If, during a series of pushes, each one is aimed to slide the rock in the same direction, it may be moved out of the area without any direct contact with the blade.

While the dozer is backing up, it can move a rock sideward by the maneuver shown in (B). When the side of the blade touches the rock, the steering clutch on the opposite side from the stone is released and the brake applied hard. The dozer will spin on the braked track, the side of the blade will push the rock, and it may move several feet. The rear of the track is sometimes used in the same manner while going forward to move a rock a short distance when restricted space makes it difficult to get at it otherwise.

Very heavy rocks, or those embedded in soil, should not be moved in this manner because of excessive strain on the tracks.

Pitching. Tracked vehicles pitch badly in walking over ridges, stones, or poles. Figure 49 shows a series of positions assumed by a dozer walking over a small log. After overbalancing, the machine generally falls with a crash which is very damaging to both tractor and operator. Such a

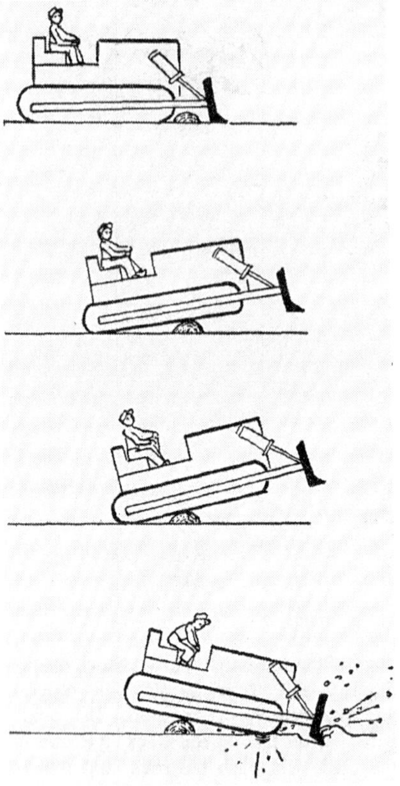

Fig. 49. Wrong way to cross a log

log or bump should be pushed out of the way or avoided if possible. Crossing, if necessary, should be done slowly and at an angle, so that one side of the machine crosses the top and starts down while the other is still climbing. This slows the fall, and avoids any danger of turning over backwards. If the bump is a soft ridge, turning the tractor sharply while crossing it will cut it down.

When a bulldozer is digging at capacity, the tracks often spin a little, then grip and move, and then spin again. Each time they spin, they build up piles of dirt at the back. If the machine backs up in the same track, these piles will have the effect of the log in the illustration, causing the back to rear up and then fall. This jolting can be avoided by keeping a wide enough work strip so that it is not necessary to use exactly the same path backward as forward. A tractor is not much affected if only one track crosses a bump.

Tracks will carry the machine over a narrow ditch without any pitching, if the movement of the tracks does not break down the banks. A pair of quite frail boards will often be sufficient to prevent caving as they will protect the bank against the backward push of the grousers. Crossing of a ditch wider than one quarter to one third of the length of the track on the ground is liable to be unsafe unless it is quite shallow.

Rear Power Control. Many dozers have pumps or winches driven from a rear power take-off, so that they do not operate when the flywheel clutch is released. This lowers the efficiency of the machine in hard or rocky digging, in uprooting stumps, in handling of bulky objects, and in many jobs where it may be desirable to stop the machine momentarily to raise or lower the blade, and then continue. The operator of a rear pump job will de-clutch, shift into neutral, adjust the blade, shift into gear, and engage the clutch. With a front pump the clutch is merely disengaged and re-engaged.

The effect of a front pump may be obtained by stopping the tractor by releasing both steering clutches instead of the flywheel clutch. This involves greater driver effort, and may result in higher maintenance because of the greater cost of the steering clutches.

Gears. Bulldozing may be done in low or second gear. Low is less taxing for the machine in heavy pushing, and easier for the operator in precise work. Second is considerably faster, and in clean material good loads can be moved and smooth grades maintained. However, the higher gear makes it harder to cope with stones, hard spots, and other difficulties. Loose gravel may float a loaded blade so that it will not dig in.

The amount of traction often determines whether a particular job can be done in second. A machine with narrow tracks, flat shoes, or worn grousers, or a standard machine on loose soil, spins the tracks easily so that when too deep a bite is taken or an obstacle is hit the load or shock to the engine is cushioned by slippage in the tracks, giving the operator time to raise the blade or disengage the clutch before stalling, even in a high gear.

Dozing Cycle. Most bulldozer digging is done in shuttle fashion with the machine facing in one direction through the dig, push, spread, and return parts of the cycle. This is because the distances covered are usually quite short and turns, particularly in soft dirt, take time and spoil the grade, so that it is quicker and easier to back to the cut than to make two turns in order to use a higher gear. On pushes of one hundred feet or longer the turns may be better unless the machine has a fast reverse.

Many tractors in the medium and small class have only one reverse, while heavy ones may have from two to six reverse speeds. A bulldozer needs several reverse speeds almost as much as it does forward ones. Backing up is the unloaded part of the cycle where lack of work should allow high speed, but if a single reverse is used it must be powerful enough to climb steep grades and pull out of mudholes, and so cannot be fast.

The backup part of the cycle may be put to work by using back-ripper teeth on the blade to loosen soil for the next push.

Speed in reverse may also be limited by the quality of grading done during the push. In making heavy cuts it may not be efficient to take time and skimp loads to make a level floor with each pass, particularly if the soil is coarse. Gouges and bumps may be left by the blade and humps made by spinning tracks. The result is that a slow return may be made to avoid pitching even when a higher gear is available.

Hill Work. Dozers may be used on moderate side slopes and wide track models on steep ones up to 20° or more. However, they are quite likely to upset unless care is used. A machine which appears to have an ample margin of safety may be suddenly flipped on its side by running over a stone with the higher track, at the same moment that the lower track enters a hollow or soft ground. This is less apt to occur if the machine is pushing rather than walking, as it will then be moving slowly, will have the blade close to the ground, and will be steadied by the load. It also obtains some support from the windrow spilling from the downhill side of the blade.

Working on frozen slopes is hazardous as the grousers may act like skates and allow the machine to slide uncontrollably downhill, regardless of the direction in which it is facing or trying to move. Sharp ice cleats will hold in such conditions but dirt grinds their points off very rapidly.

A similar danger is encountered on rock slopes, particularly shale with beds parallel to the surface.

Slopes on soft fills are very treacherous, as the tip will be increased by the lower track sinking deeper than the upper one.

If a machine starts to roll over slowly, it can sometimes be saved by turning downhill and lowering the blade.

A slope which is too steep to be safely worked sideward may sometimes be graded by running the dozer along it diagonally. If it is too steep for this, soil may be pushed straight down from the high spots, moved along the bottom, then pushed up to the low places.

Dozers can safely negotiate very steep up and down grades. Digging and pushing efficiency are much greater downhill and taper off to zero on steep upgrades. Steering is apt to be tricky on steep slopes, whether up or down, because of track slippage and shift in the center of gravity. Very steep grades of 25° or more should be climbed forward rather than in reverse, because of better balance and traction.

Cutting should be done downhill whenever possible and in very hard ground it may be advisable to dig it downhill, even if the spoil must then be pushed up the same hill for disposal.

The engine oil pressure gauge should be watched closely on steep work as some engines do not get proper lubrication when tilted steeply, especially at compound angles, and a low oil level which still gives adequate lubrication on a level may leave the pump dry on either up or down grades.

Where a run includes a down slope the operator may push several loads to the top, then push most of the resulting pile down in a single pass.

Unless the ground is loose a dozer can push much more of it than it can cut and move while cutting. Here again it may be good technique to drop one or more loads at the end of the cut, pushing the final load all the way through, along with the bulk of dirt piled previously.

Cable Operation. There are a number of differences in operation and performance between the hydraulic and cable controls. Cable usually moves the blade somewhat faster, the lift having a speed of a foot a second or better, and the drop is that of a free falling body. If a slower lift is desired, it is usually made in a series of short jerks, although some cable controls allow partial engagement of the clutch for a slow smooth rise.

The blade may be lowered by allowing it to fall, but it is desirable to check it with the brake just before it touches the ground to ease the shock, and it is necessary to have the brake applied just after it hits so that the drum will not continue to revolve, unspooling and snarling the cable. A smoother technique is to let the brake drag slightly as the blade goes down, or to alternately release and apply it (pump it down).

The depth of cut is regulated by the cable, but control is not as complete as in the hydraulics. The blade cannot be forced down by the weight of the tractor, so it tends to ride up over hard surfaces. If it uses a sharper cutting front, the blade is more apt to be sucked down into plastic soils, and being suspended from the radiator guard can quite readily pull down by compressing the tractor springs, and so will cut too deep unless carefully controlled. Engine torque partly counteracts the pull-down.

Digging may be done with a few inches of slack in the cable allowing the blade to find its own depth, or with a taut cable. Grading is done with a tight cable. It is bad operation to let the cable get very slack as it snarls and knots, shortening its life, the excess may catch on objects, and lifting of the blade is delayed until the slack is reeled in.

Cutting Hard Ground. If the blade refuses to cut down, it should usually be tipped forward by means of the pitch rods. This helps it to cut when it is empty, but a load can float it out of the ground more readily. If it will not cut into humps or a bank, which is less usual, it should be tipped back. For general pushing the blade should be centered or back, as it rolls the material most effectively in that position, reducing friction. See Figure 50.

If the blade will not cut after adjusting the pitch, a limited amount of ground can sometimes be cut up by spinning the tractor, first on one track then the other, on the area to be dug. The grousers may chew up the ground sufficiently so that it can be bladed off readily.

Digging in hard or stony ground will

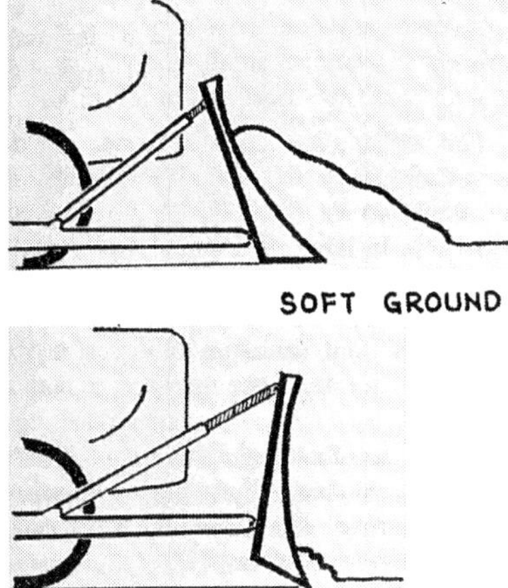

Fig. 50. Dozer blade pitch

be easier if the work is arranged so that cuts are made by only part of the blade. One corner of the blade should be tilted down if possible. If not, one cut can be ground down to a depth of a few inches and overlapping cuts made at the sides. If half the blade is in the air over the cut, weight per inch of edge will be doubled for the part on the ground and it will cut more effectively. The dirt it gets under will add weight to the blade so that it will probably do some digging in the original cut in addition to removing a substantial bit at the side. Effective work may also be done by working outward from a cut, herringbone fashion.

If any considerable quantity of hard dirt must be bulldozed, it should be loosened by back ripper teeth, by a separate ripper, or dug with hydraulic controlled dozer shovels or dozers.

Because of the usually unlimited drop of the blade below the tracks, cable dozers are able to cross sharp ridge tops without losing blade contact with the ground and are therefore preferred in cutting roads or firebreaks across rough country.

Comparison of Hydraulic and Cable Control. Cable bulldozers have several advantages over the hydraulics. They have faster lift and drop, and start moving with a yank which is very effective at breaking out. The blade can be dropped far below the tracks and usually has a high lift also, but it cannot be forced down into the digging by the weight of the tractor. The sharper digging angle often given to the edge and lower moldboard to compensate for this makes dumping loads difficult on steep upgrades, as the dirt rests on the raised blade.

On the maintenance side, cables need periodic replacement. The cost of the cable is not large but the nuisance of having it on the job and installing it during work time is a factor to consider. More lubrication points are present, and sheave bushings and clutch and brake linings need occasional replacement. The dozer is useless for lifting the machine out of mud.

Hydraulic systems have no fast wearing part to correspond to the cable, but packing glands need regular inspection, with adjustment and replacement; other leaks in the system must be watched for; breakage of flexible hose or pipes can put it out of action, and pumps, rams and valves need occasional replacement or repair. Oil must be added to the system rather frequently and changed occasionally. Operation is slowed by cold.

The decision about which type to use depends on the work and on the size of the machine. If the digging is such that a cable blade has difficulty cutting it and a hydraulic digs it easily, a hydraulic should be used. If it is so hard that the hydraulic has difficulty, it should be broken up with rippers, after which either can handle it easily. Stumps in soft ground respond best to the yank of the cable, but in hard ground down pressure is needed to get a standard blade under them. Results in fine grading depend more on the particular machine and the operator than on the type of control.

Cable blades give much better penetration in large sizes than in small. The smallest dozer blade is about five feet wide and weighs two or three hundred pounds, while the largest mounted on a crawler tractor is about thirteen feet wide and weighs several tons. It will be seen that the weight on each foot of edge is much greater in large machines than in small, and it is this which largely determines whether a particular blade edge will penetrate. The largest machines have a penetration sufficient for most soils, and although hydraulic dozers of the same size can put several times as much weight on the knife, this extra is less often needed.

Hydraulic blades are therefore usually preferred in medium and small dozers, and cable in large units. This is not a hard and fast rule, as even a large dozer purchased for work in hard ground, in soft mud, or at miscellaneous work should be hydraulic, where a small one to be used only in pushing loose material might as well be cable.

For much the same reasons, towed accessories such as scrapers and rippers are liable to be hydraulic in small sizes, and cable in large. It follows that a cable unit for a large dozer and a hydraulic one for a small are more liable to be useful with accessories, and therefore more useful to the contractor than the opposite arrangement.

A curious feature of the controversy between cable and hydraulic enthusiasts is that some of the important differences are not made necessary by the type of control but by tradition. A cable blade can have down pressure—as evidenced by a special unit supplied by one dozer manufacturer. Hydraulic blades can be made faster than cable, by enlarging pump and lines, or increasing the pressure, and could be dropped to any depth desired at additional expense, by alteration of rams and levers.

CHAPTER FOUR

TRACTOR LOADERS

DOZER SHOVELING. Shovel dozer attachments consist of a push or lift frame and a bucket which can be raised, lowered, and dumped hydraulically or mechanically. They are usually mounted on wide model crawler tractors, with special long track frames that have an extra truck roller on each side and heavy duty idlers.

These machines lift and lower the bucket by means of two-way rams acting against the bottoms of the lift arms, and adjust its cutting angle and dump it by means of another pair which act through levers against the upper part of the bucket. See Figure 51.

Bucket Angle. When the dump rams are fully retracted the standard bucket floor may be level or have a slight downward tilt, and the edge be in line with the push arms. This position, called flat or closed, puts the least strain on the bucket, and the dump arms and hydraulic system, and gives efficient penetration into banks and high spots. It is the best angle for raising and carrying a load.

For cutting down into hard dirt, the bucket should be tipped downward ten to thirty degrees. When it has penetrated to a depth of two to six inches it should be flattened all the way while the forward motion of the tractor is continued until the bucket is filled. This combines good original penetration, sturdiest position of the bucket for most of the pass, and a powerful prying effect during the change in angle. Under some soil conditions continual minor adjustment of the angle while digging helps penetration.

The flat position is best for pushing a quantity of loose dirt, but the bucket should be turned down steeply for spreading and grading it so that dirt will flow freely off the floor into holes, and so it will not be pulled down by sticky soil. Care should be taken not to hook into solid obstructions at a steep angle, as the bucket is then in its weakest position, and leverage against the dump mechanism is at its maximum.

The depth of cut may be regulated either by the hoist or the dump rams. For any given position of the push arms the edge will be highest when flat and two or more feet lower when fully dumped.

Digging Banks. The bucket has much weaker penetration in proportion to output than a dipper stick because it is larger in relation to the power and weight of the machine, has a wider edge in the digging, and may lack teeth. Another digging difficulty is that the hoist is very slow in proportion to the speed of the tractor, so that the bucket tends to get under more dirt than it can break loose and lift. The effect is of a dipper stick shovel with no retract and with the ratio between crowd and hoist fixed so that the bucket always tends to crowd too deeply into a bank.

If the bank is of hard material it is desirable to keep it sloped back to facilitate

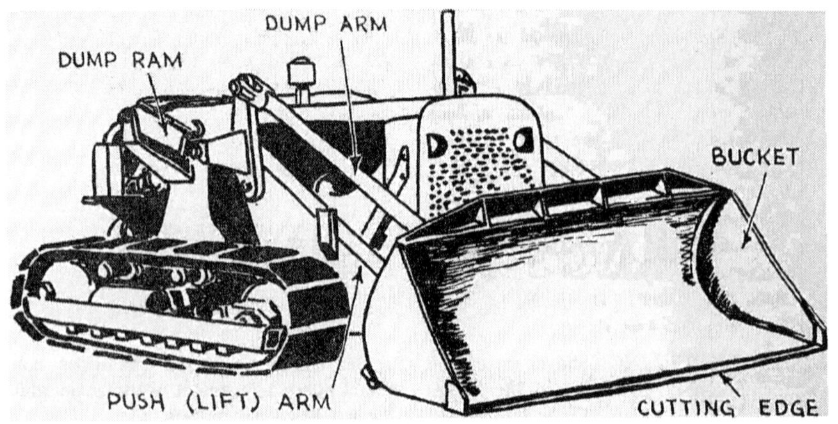

Fig. 51. Dozer shovel

breaking out the loads. This may be done by digging in layers down from the top of the bank, then slicing upward from the bottom. The slope may be maintained by making each cut parallel with it. It takes a few seconds longer in each cycle to dig this way because of the extra distance traveled, but it may be easier for the machine than digging in chunks.

A more usual method is to force the bucket into the toe of the bank and lift it. When the tractor lugs down the main clutch is disengaged. The engine recovers its speed and the bucket usually will rise, cutting the bank and partly filling. When it is up a few feet, the clutch is engaged again, crowding the rising bucket into a higher part of the bank. This process is repeated until the bucket is full or is above the bank. It is sometimes helpful to crowd the filled bucket against the upper part of the bank, as this may break it up so that it will slide down to the toe where it can be picked up easily on the next pass.

The engine clutch may also be slipped to hold the bucket in the bank enough to fill without crowding it in so hard that it cannot rise. These methods, particularly the second one, will increase clutch wear, but if this unit is easily adjusted and replaced the greater working efficiency will justify the extra maintenance.

If the bank is only two or three feet high, it may be dug by keeping the bucket at final grade and running beside the bank, cutting into it as much as possible without slowing the tractor excessively. The cutting will be done by the side and floor of the bucket which is in the bank. The soil will roll along the bucket and fill it, although the load will be heavier on the bank side.

If the machine is driven head-on into the bank and gets under too great a load, the bucket may remain stuck and the back of the tractor rise. In this case the tractor should be shifted into reverse and backed out. As soon as the bucket is pulled back slightly in the bank, it will rise and usually pull out with a good load.

Payload. The amount picked up in the bucket varies with the nature of the material, slope of the bank, the surface to be walked over, and the skill of the operator. A standard one yard bucket load will vary from one-half to one and a half yards, and should average seven eighths of a yard to a yard in medium digging.

When dumping close to the digging point, as in side casting, or loading a properly placed truck, keeping up a fast cycle is usually more important than getting maximum loads with each pass. As distance to the dump point increases, capacity loads become more important than the time taken to get them.

If the load must be carried over rough ground or backward up a slope, it should be limited to the weight the machine can carry easily without tipping.

If the bucket does not fill sufficiently the tractor should be backed, the bucket lowered to floor level, and another pass made. If the load is one sided the second cut should be made at an angle to the bank so that the empty side will penetrate first.

It is often difficult for the operator to judge the amount in the bucket unless the digging is easy enough to permit dirt to be forced over the top. It is a good idea to cut a row of holes, about an inch and a half in diameter, in the upper back of the bucket, as in Figure 52, through which the operator can see whether it is filled. Spillage through such holes is negligible in ordinary digging.

This figure also shows the construction of grab brackets for convenient attachment of chains.

Ramping Down. If the digging is downward, as in cutting a cellar or a ramp, hard material may require pitching the bucket floor at a twenty to thirty degree angle to the line of the tracks, and cutting in thin slices. The downward pitch of the ramp should be gradual, as the machine is nose heavy with a loaded bucket and may tip over frontward in backing out of the hole.

Gouging. A common difficulty in digging heavy soils is that the penetration of the bucket is too good so that it will be pulled down by the slice it has dug, either raising the back of the tractor or pulling the front of the tracks down into the ground. This may be combatted by keeping the bucket as flat as possible, or by tipping the floor into a nearly vertical position which cures the difficulty but puts extra strain on the bucket.

It is often best to let it gouge, and make

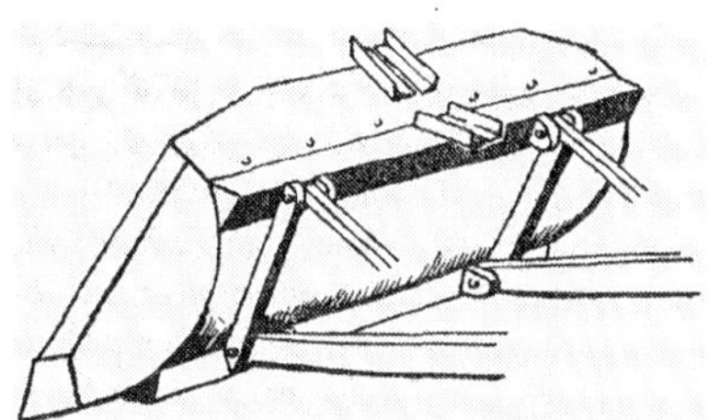

Fig. 52. Bucket, and chain grabs

an extra pass to grade off the area as often as necessary.

Transporting. If the ground between digging and dumping is hard and smooth, backing, turning, and dumping can all be done at speed with safety. If the ground is rough the machine must move slowly, as going over a bump or ridge with a heavy load may cause it to fall forward, the bucket dropping to the ground and the operator's seat rising into the air. If part of the load dumps, the tractor will settle back. In crossing rough ground the operator should be alert to lower or dump the bucket if an upset starts. Lowering has the effect of taking the bucket weight off the tractor long enough to enable it to recover its balance.

Dropping a loaded bucket and stopping it abruptly in the air may burst a hoist ram hose if it does not overbalance the machine.

It is advisable to cross ridges at such an angle that one track will be part way across before the other reaches it. If the ridge is soft it may be possible to cut a more level path through it by turning sharply while on it.

If the bucket is carried two to four feet above the ground the consequences of overbalancing are unlikely to be serious. If it is high the weight is not quite so far forward, so tipping is less likely; but if it occurs, it will give the operator a worse toss. A high held bucket also involves the danger of less likely but far more serious side tipping.

The bucket should not be given a high lift if there are rocks or lumps projecting over the back, as they might fall and injure the operator or the tractor. Normally such things fall to the ground, or on the front of the tractor without serious damage, but they might roll down the arms or bounce back off the hood.

TRUCK LOADING. Maneuvering the Loader. The actions involved in loading a truck are as follows: The tractor is headed toward the pile or bank, is put in low gear, the throttle is opened, the dump cylinders are retracted against their stops, and the bucket is lowered until it almost touches the ground. The clutch is then engaged. As the bucket enters the bank the hoist lever is moved to UP position so that the bucket will rise as it penetrates. When it is filled, the main clutch is released, the gear changed to reverse, the clutch engaged, and the machine backed out. The hoist lever may be moved to HOLD when the bucket is a few feet above the ground.

While backing, the tractor should be turned so that it faces toward the side of the truck to be loaded. It is then shifted into low or second, the hoist lever moved to UP, and the machine walked forward. The bucket should be lifted high enough so that the downward movement of the bucket lip during dumping will not cause it to strike the truck, and it is good practice to have it high enough to clear the side to avoid accident while backing. The hoist is usually completed before the truck is reached. The control is then moved to HOLD and the tractor moved so that the bucket is over the truck body.

A good procedure is to time the lift so that the bucket will just safely clear the truck as it is moved over it and the control can be left in UP during the dump. When the dump valve is opened it blocks the hoist valve so that lifting stops, to recommence immediately when the dump valve is closed. This "live" bucket is more easily controlled over the truck body than when the hoist is in HOLD.

In either case, the tractor is walked forward until the bucket is as far over the body as desired or until the radiator guard touches the truck body or tire. The main clutch is disengaged, the tractor held with a footbrake, and the dump lever moved forward to dump the load. The first bucket or two are best dumped slowly to reduce shock to the truck. If the soil is sticky the bucket may be shaken by banging against the dump stops by moving the dump valve lever rapidly back and forth.

The tractor is placed in reverse and backed away, the bucket lip being raised to clear the body if necessary. When clear of the truck, the machine is stopped, put in low or second, and headed toward the bank, the bucket being put in digging position and lowered during the return trip.

This loading cycle will vary between thirty and fifty seconds under average good conditions. Easy digging, proper truck spotting, and low bodies favor a fast cycle.

Spotting Trucks. There are many possible patterns of digging and dumping, some of which are shown in Figure 53. (A) is the most used method, in which the side of the truck is at right angles to the face of the bank. This involves a quarter turn twice in each digging cycle. The turn may be difficult in soft ground and the tracks are subjected to considerable turning stress.

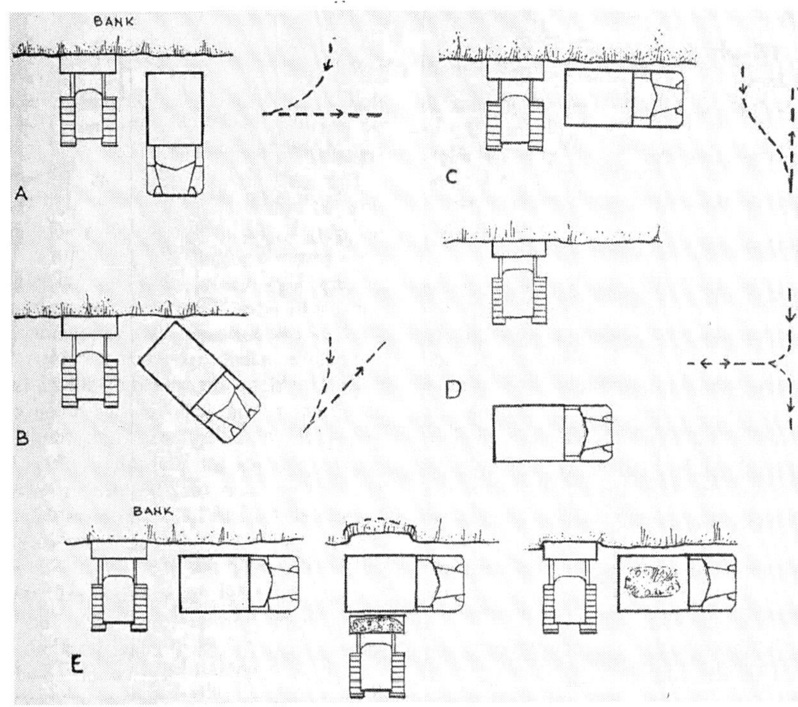

Fig. 53. Loading with dozer shovel

In (B) the truck is parked at an angle of about 45° to the bank, so the loader's turns are only half as sharp. In this position some loads can be swung from the bank onto the truck with a very short backward movement, thus increasing loading speed. This is the best system.

In (C) the truck is parked parallel with the bank and the digging is done just behind it. This involves about the same amount of turning as (B) with a greater amount of walking.

In (D) the truck is parallel to the bank but at a distance from it, so that the tractor must make a half turn with each load in order to get it in the truck, and another half turn to get back to the bank. More walking is required than in other methods. This is the slowest and most unsatisfactory of the arrangements suggested, but is often used in muddy or sandy pits where trucks are restricted to certain drives.

In (E) the truck is again parallel to the bank, but it comes to the loader for each bucketful. The machine fills its bucket from the bank and backs straight away. The truck backs in front of it, the tractor advances and dumps into the truck, the truck moves forward, the loader takes another bite, backs up, and waits for the truck to come back.

This method involves no turning, is fast and efficient if properly coordinated, and keeps wear on the tractor to a minimum. However, it is disliked by truckdrivers and is not particularly recommended for that reason.

Loader output can usually be increased substantially by having a spotter place the trucks, or by training the drivers to be alert to the needs of the machine so that they will not only take a convenient position, but be ready to move if the machine works away from them. This last is particularly important where the digging is shallow and truck bodies large.

Dumping in Body. The width of the standard yard bucket is about six and a quarter feet, and the truck bodies to be filled are usually between seven and eleven feet long. A short body can be given a full heaped load by dumping in the center only. A long body can be given a good load by dumping alternately in the front and rear from the side, but there is a tendency to pile up in the center and skimp the corners. If a very heavy load is desired it may be best to load the front of the body from the side, then load the rear from the rear.

Clearance under the conventional one yard bucket edge in fully dumped position is over seven feet, but the machine is at its best in loading trucks with sideboard heights of six feet or less. At this height the loader dumps about three feet ahead of the radiator guard, which makes it possible to slide the load to the far side of the truck if necessary, and the operator can see into the body. With increase in lift the bucket moves back toward the tractor and loses reach, so that a truck with sides seven feet high would require extra work moving the load across the body or loading from both sides.

The larger loaders have sufficient bucket width, dumping height, and reach to load the largest bodies readily without the procedures described below.

High Trucks. If a body too high for convenience is loaded, a heaping pile of soil is built up on one side. This heap may be moved toward the other side by pushing it with a loaded bucket held with its floor parallel with the ground and just clearing the sideboards. The bucket is then dumped and its load pushed over by the next bucketful, this process being repeated until the body is full.

Dirt may also be worked over by putting the bucket in fully dumped position, dropping it just inside the body and rotating it part way toward flat position.

Very high trucks may be loaded by using both sides. The bucket must be rotated to flat position after each dump in order to be pulled back over the side.

If many big trucks are to be filled it may save time to dig slots two to three feet deep into which they can back while the tractor operates on the higher level. If the pit floor must not be torn up, a few bucketsful will build a ramp up which the tractor can walk to get an easy working height. The ramp should be made so that the machine is not tipped up steeply while dumping, as this reduces reach and increases the effort of holding position.

If there is uneven distribution of earth front and rear while loading, it may be corrected by cutting into the bank at an angle so that the side of the bucket which will dump over the low spot will get the heaviest part of the load. Dirt already dumped can sometimes be rearranged with the bucket in dumped position by pushing, pulling, or turning.

Distances. The loading is fastest if the truck is so close to the digging that the machine has just comfortable room to turn, although the hoist is not fast enough to lift the load to the required height in a very short travel distance. With steadily diminishing returns on a yards-per-hour basis the dozer shovel can load at considerable distances. When a pit is too wet or sandy for truck operation it can dig in the pit, carry material onto firm ground, and there put it in the truck.

Operation in digging in one place and dumping or stock piling in another is similar to truck loading if it is desirable that the space between the digging and the dumping should not be disturbed. If the distance is short, 50 feet or less, or a big yardage is to be moved it is usually more economical to push the material with a flat bucket, as the machine will push from two to three times as much dirt as it will carry, which on a short run will more than compensate for slower speed.

Carrying may be done in third to top gear on smooth ground, while pushing requires low or second. If the tractor has only a single reverse its return trip is slow in either case, and on a long carry it will pay to turn it around for a fast run back to the pit. The return trip may also be used occasionally to level the ground, either by pushing or back dragging.

Output. A crawler mounted front loader with a full size bucket may be expected to load dirt into properly placed trucks about as rapidly as a dipper shovel with half its bucket capacity. That is, a one yard dozer shovel is roughly equivalent to a half yard shovel in medium digging.

The loader will show a proportionately greater output in softer dirt which heaps on the bucket, provided the pit floor is firm.

The dipper shovel will show a smaller decline in production as the digging becomes harder or coarser, and it is less bothered by rough, rocky, or soft pit floors. However, the loader will pick up larger rocks without chaining.

The loader can move around more readily, can keep a smooth floor and clean out boulders without assistance, and can be used as a dozer when not needed for loading. Its maintenance cost will be higher because of track and roller wear.

HANDLING OVERSIZE. This bucket is not well suited to picking up bulky objects because the floor lacks depth and at ground level it may incline slightly downward. The overhang of the back makes it hard to balance anything big on the floor.

If stumps, boulders, or other large pieces are to be picked up, they can often be crowded against a steep bank which will prevent them from falling out while the bucket is lifted high enough so that the floor will slope back. Two loaders can lift the object between them until it settles into one of the buckets.

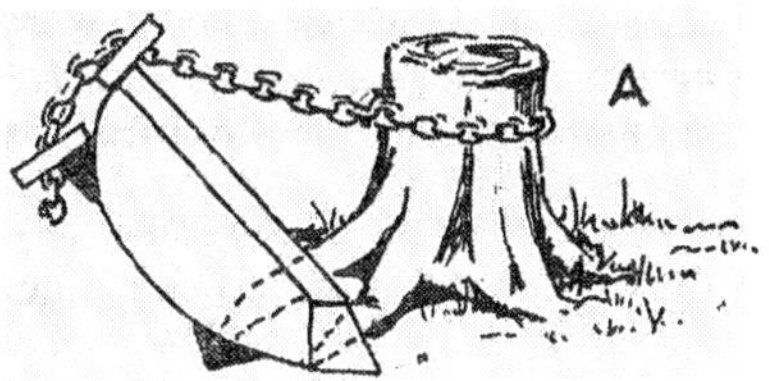

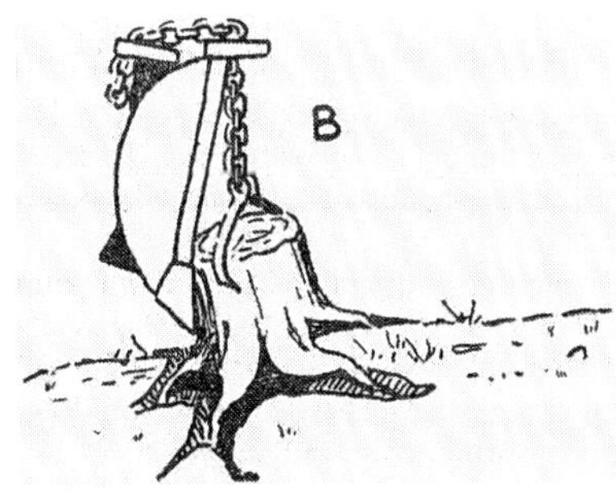

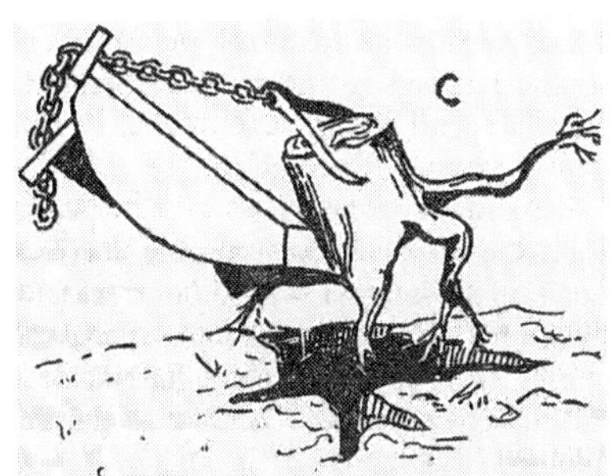

Fig. 54. Lifting stumps

If the object is too large to be picked up in this way or there is nothing to crowd against, it can be maneuvered onto the floor and held from falling out by a chain to the top back of the bucket, as in Figure 54 (A). This method is particularly effective with loose stumps resting upright.

A load fastened in this way usually settles as it is raised so that it is easy to unhitch before dumping. If it settles forward and puts so much tension on the chain that it cannot be released, its overhang may be lowered so as to be supported by the truck sideboard or the pile. The chain may then be removed and the load boosted into position.

The bucket may also be lowered over a loose stump in dumped position, and attached by a chain or chain and tongs, as in (B). The bucket is raised and flattened and the stump is pried up by the edge to nearly floor level, as in (C).

For efficient application of these methods, or for crane work in general, it is almost essential to have a chain grab bracket built onto the top of the bucket. For use in dumped position, a second bracket on the rear beam is desirable.

A chain should not be anchored on a push arm and passed over the top of the bucket, as these change their relative positions during a lift, and the chain is likely to be stretched and broken.

Crane Work. The machine may also be used handily as a crane by means of a chain fastened to the back of the bucket and dropped across the edge. Height of the load may be regulated both by the hoist and dump controls.

If the lift is not high enough to put the load on a truck, a small log or a couple of rocks may be placed where the tracks will climb them as the machine approaches the truck, which will cause the tractor to rear up and increase the load height substantially.

The crane attachment for use in place of the bucket has a somewhat higher lift and is more convenient to use than the chain and bucket, but for incidental work it does not justify its cost or the time required to change over.

If the bucket is placed in a fully dumped position, a beam or log may be wedged or lashed to a dump arm so that it lies on the bucket and extends forward from it. This will be raised by lifting the push arms or flattening the bucket, and will give a very high lift to light loads.

Valve action is not smooth enough for precision crane work, but the mobility of the machine and its ability to work on soft footing and in restricted spaces qualify it for many hoisting and moving jobs.

The loader may also be used to carry or raise material piled in the bucket by hand. This "motorized wheelbarrow" work may be a profitable sideline when grading around buildings under construction.

BULLDOZER BLADE. A bulldozer blade may be substituted for the bucket. Its use is advisable if the machine is to be used as a pusher for scrapers, or if it is desired to equip it with grousers and do heavy digging in rocky soil. The blade is more rugged than the bucket. A wide blade enables the machine to turn in its path without climbing onto the edges of the cut. However, the standard width bucket will do most bulldozing jobs better than a bulldozer, and is immediately available for loading or other special work.

This blade is somewhat more effective than on a regular bulldozer because the digging angle can be changed with the dump control to meet any change in the digging, and because the blade can be dumped at the top of steep piles, where part of the load has a tendency to ride back down the pile on a standard blade. In addition, it can be floated forward in full dump position as well as backward to smooth over loose soil.

Bulldozing with Bucket. The standard bucket may be used as a bulldozer by cutting with it in a digging position and spreading with it more or less dumped. In the latter position the bucket floor will have a tilt smiliar to that of a blade. Its advantages over a blade are the greatly increased cutting efficiency due to the knifelike edge; its ability to push bigger loads of soil because a good pushing load is contained in the bucket where it causes so little load or friction that nearly a full blade load can be pushed ahead of the bucket; faster grading work because of greater transporting capacity and ability to carry dirt to hollows without disturbing surfaces in between; the facility with which material can be backed out of banks, and the exactness with which it can be deposited where needed.

This machine has a greater effective reach below the tracks than any bulldozer. In the vertical position the bucket edge will reach more than two feet below the tracks, and it will cut a sheer wall to that depth. Because of their wide bases no dozer blades can be forced straight down or reach such a depth in a short distance, even when capable of such a drop.

This depth ordinarily cannot be reached in one cut, but by several with the blade at a successively steeper angle, or the push arms lower each time. In itself it is only occasionally useful in ditching or squaring off sides of shallow excavations, as the tractor cannot follow the bucket over such an edge without special precautions.

However, this ability to reach down enables the dozer shovel to cross sharp ridges without the bucket losing contact with the ground, a matter of great importance in clearing work. The bucket may also be used to ease the machine down lesser drops by holding it a foot or so above the bottom and walking very slowly off the edge. When the tractor over-balances it will be supported by the bucket, which can be gradually hoisted, thus lowering the front of the tracks to the bottom. This process may crush the bank enough so that the machine can then be walked off it, or it may be necessary to put small blocks to reduce the drop of the back.

The bucket is extremely efficient at backdragging because the floor, when tilted straight down, can penetrate fairly hard material vertically and the clear space between the bucket and the tracks is sufficient to accommodate a lot of dirt. When backdragging in order to smooth loose dirt, the operator has a choice of tilting it to full dump, where it will cut, through intermediate positions where it will not cut and will pull decreasing quantities of earth, to flat where it will move hardly any. In the flat or near flat positions, down pressure can be applied so that loose dirt may be compacted.

The transition from forward digging to spreading is most easily made by putting the lift lever at UP, and dumping the bucket a little at a time by rapid back and forth movements of the dump lever. Each dumping movement should lower the edge just enough to compensate for the amount it was hoisted while the dump lever was in neutral. As the hoist will not operate while the dump ram is moving, it is possible after practice to keep the edge running an even grade this way. The dumping operation may be stopped as soon as the floor has enough tilt to spill the entire load. Hoisting may be interrupted occasionally if a thin layer is wanted.

If the dirt is to be placed in a pile, the same method may be used, starting the dump with the edge kept down soon enough so that the bucket will be in dumping position at the pile. The bucket can be allowed to rise while the shift into reverse is made; or can be left stationary during the shift and lifted slightly while backing out.

Another system is to leave the bucket in pushing position until the place to dump is reached, putting the hoist in UP, shifting into reverse, then dumping and backing away.

If a high pile is being built, either method may be used, simply lifting the bucket to conform to the slope of the pile while climbing it. It is a good plan to knock the top off the pile occasionally, by forcing the bucket down in dumped position in the loose dirt a few feet short of the top. Several yards may be pushed over at a time this way, shortening the climb to dump the next few buckets.

General Considerations. It will be apparent that the operator must use two controls more or less continuously, as against one in a bulldozer, in order to bulldoze with a bucket. Some operators may dislike the machine for this reason.

Another drawback is that location of the push arms between the tracks and the narrow path cut by the bucket enable stones and dirt to roll between the bucket and the track, making it necessary to lift the bucket and back far enough to get behind them in order to get through or to keep a grade. Also, the top of the bucket may be in the way of pushing close to a building.

A more serious consideration is the possibility of breaking up the bucket. It is not practical to make it as heavy or as strong as a bulldozer blade, and maintenance costs in heavy digging, particularly among rocks, may be higher. This is especially true if the operator does not keep the bucket at a proper angle when digging or if full grousers are used on the track shoes.

The use of the heavily built rock bucket for hard digging will materially reduce maintenance costs, but it necessitates the lifting of several hundred pounds of extra weight when loading.

The dumping rams may be used to push or pull the tractor when lack of traction or breakdown prevent it from walking in the regular way. The bucket is pushed down in firm ground or on a mat of poles or planks. Dumping the bucket will pull the tractor forward, closing it push back.

This control over the bucket edge can also be used in pulling stones back out of water or marsh, and in reducing the length of the machine to fit it in a short space.

SKID SHOVEL SHOES. An interesting feature of the skid loader is the use of ground shoes under the push arms at the bucket hinge points. They serve as fulcrums for prying action in breaking out hard digging, as bases from which depth of digging can be controlled, and as skids for transporting loads. With shoes resting on the ground, the dump rams of the 1¼ yard model can exert an upward force of almost six tons at the bucket edge, without putting any lift load on the tractor, as against the two ton lift offered by the hoist rams.

Digging depth is controlled by resting the shoes on the ground and rotating the bucket forward or backward for greater or less cut below the shoes. An indicator on the left dump ram shows the depth of this cut in inches.

If the indicator is set at zero the bucket edge is in line with the bottom of the shoes and will not cut down, but will shave off bumps or cut into banks. If it is set for a two inch cut the edge will dig in that distance as the tractor moves forward, and on a level grade the slice removed will be two inches thick. However, when the shoes slide into the cut the depth must be reduced to zero if a two inch cut is to be maintained, as the cut is figured from the bottom of the shoes and not from the original grade.

Such digging can ordinarily be done in FLOAT position. If the ground is hard or stony down pressure may be required to keep the edge from riding up and carrying the shoes clear off the ground. When the ground is soft allowance must be made for the shoes sinking in as this will cause a deeper cut than that shown by the indicator.

When this machine is fitted with a dozer blade, depth of cut is regulated in the same manner by tilting the blade edge below the shoes.

When the bucket is filled or the back of the cut is reached, the bucket lip should be tilted sharply upward before raising. This serves to pry out any hard material over the bucket, puts the lip at a better angle to penetrate as it rises, and reduces spilling of earth off the front.

If the load is heavy, must be carried a considerable distance, and the ground is hard, the shoes may be left floating on the ground and the load held in the tipped back bucket and skidded to its destination, saving wear on the idlers and tracks at the expense of overcoming some extra friction.

In other respects, operation conforms to the general pattern described for crawler-mounted loaders.

Gravity dump bucket. A gravity dump bucket which is optional equipment on some makes of loaders, is so hinged to the push arms that its weight is to the front, and it will dump almost instantaneously when the latch is released by means of a cable and hand lever. It is stopped at the end of its swing by a pair of coil springs that serve the double purpose of cushioning the bucket against slamming its stops and giving it a vigorous shake which helps to dislodge sticking material.

After dumping the bucket is returned to digging angle by lowering its front onto the truck side or the pile, or dragging it backward on the ground, so that it is pushed up until the spring latch snaps in place and holds it.

Digging angle is set by means of a pair of threaded adjustments with a rather small range.

This type of bucket is less expensive than the hydraulic dump type, but it is not as effective for general work. The lack of quick control over the bucket angle hampers penetration. Grading and bulldozing may be done by cutting in digging position, and spreading and grading fully dumped, but this is not entirely satisfactory and a blade should be installed if much dozing is to be done.

The latch dump bucket load falls in a chunk instead of pouring, which is rough on truck body and springs and shakes up the tractor. The most convenient way to relatch is to drag it on the sideboard of the truck, which will cause damage unless special protection such as heavy pipe welded above the sideboards is installed.

A number of light loaders on wheel tractors are equipped with single acting hoist rams and latched gravity dump buckets. These buckets differ from that just described in that the hinges are on the sides rather than at the back.

This bucket has the same disadvantages listed earlier and in addition may get a rock or other heavy object lodged so far back that it will not dump. Such an object may sometimes be dislodged by moving the tractor forward and stopping it abruptly with the latch released. It is sometimes convenient to have a man standing on the truck relatch a light bucket.

Some of these buckets have no adjustment for digging pitch, while others may be changed by moving the latch. It may be possible to lock the bucket in dumped position and use it as a bulldozer. This lock may also permit the operator to control the bucket swing after dumping so as to relatch it in the air.

WHEEL TRACTOR LOADERS. A special problem in using a front loader on a rear drive wheel tractor is lack of traction when carrying a load. The front axle acts as a balance point, and any weight ahead of it will counterbalance the weight on the rear wheels, sometimes to the extent of raising them off the ground. Any reduction in weight on the driving wheels reduces their traction, which in any case is not as good backing as going forward because of reaction from driving torque.

As a result of these factors, the machines cannot carry good loads on loose, slippery, or soft ground, especially in reverse. This difficulty may be reduced by heavy counterweights on the rear wheels or on the back of the tractor, by using dual wheels, or by attachment of a rear mounted tool such as a ripper, a winch, or a scraper blade.

If the load is carried very high, more of the weight will be on the rear wheels although danger of side tipping is increased. Extra traction may be obtained for a moment by lifting the load high and letting it drop. While it is falling the tractor is almost free of its weight, and the wheels may grip enough to get the machine moving. Sometimes the machine will be able to drag the loaded bucket backward on the ground or on skids.

Another consideration is that the front wheels normally carry less than half the tractor weight, but when the load is heavy they carry most of the tractor and all of the load. This results in hard turning, particularly with large machines, in which power steering may be necessary. Front tires must be heavier than the standard tractor equipment.

Most wheel tractor shovels do not have down pressure on the bucket, and there is considerable argument as to whether this feature is desirable. The type of work is the most important factor in deciding. Handling loose materials in yards and warehouses does not require it. Hydraulic dump buckets cut so effectively that they will handle many types of digging without extra weight. Wheel tractors often rear up when the edge hits an obstruction, which puts the tractor weight on the bucket without use of hydraulic down pressure.

But if the digging is hard enough to resist the edge and smooth enough not to stop it, the bucket may slide without cutting, where with down pressure it would dig right in. This occurs not only in primary excavation, but in piles which have become compacted or frozen.

Down pressure is needed more frequently with gravity dump buckets and particularly with bulldozer blades, because of their weaker penetration.

There are no disadvantages to down pressure as long as the machine is built well enough to take the extra punishment, and for general work it is apt to add enough to the usefulness of the loader to justify the moderate additional cost.

Pit Layout. The same patterns may be followed for loading trucks as with the crawler mounted loaders, except that a greater distance is traveled in order to avoid sharp turns. The wheel mounting is faster but harder to steer when carrying a load.

A wheel tractor depends largely on its momentum to drive the bucket deeply enough into the pile to pick up a good load. The speed required may be greater than would be safe for turning and dumping, and it is therefore desirable to have a foot accelerator by which the operator can speed it up as it approaches the pile without changing the hand throttle adjustment.

The pit must be arranged so that the machine will not have to carry a load while it backs uphill or across rough or sandy ground, as it is likely to lose time or be unable to work due to poor traction.

Direct digging of hard or firm soil can be made much easier by breaking up the ground with a ripper or subsoil plow, and cutting only to the depth the tool penetrates. Such a machine may be mounted on the rear of the tractor and serve for counterweight, or be a separate unit attached and pulled by the drawbar and detached during digging.

The bucket is usually a little narrower than the front wheel track, which in turn is narrower than that made by the rear wheels. This hampers the machine in side cutting banks as it is necessary to cut in on a curve, turning away from the bank side enough so that the wheels can travel in the bucket path.

Grading. Grading is somewhat difficult because the tractor cannot follow through on the bucket path. If the cuts and fills are shallow and the ground not very rough, it may be possible to continue the grade with the wheels riding to each side. More often a series of short pushes overlapping on the sides are made. In either case, if much grading or backfilling is to be done it may pay to substitute a bulldozer blade for the bucket.

It may also be possible to install a grading blade with a hydraulic lift on the rear of the tractor. This could take care of smoothing off the pit and serve as counterweight also.

Operation of front end loaders mounted on graders or trucks follows the same patterns as with wheel tractors, but much more space is required for turning and heavy loads have less effect on traction. These are mostly used in highway work in picking up windrows of surplus material, but may also give a good account of themselves in a large pit.

Dump trucks may also have a type of overhead loader which digs in front and dumps in the body.

LOADING OVERHEAD. The standard overhead shovel or loader digs at the front, lifts the bucket, and dumps it to the rear. There are also other types, one of which can dig at either end but dumps only at the front.

The digging pattern of these machines is the shuttle type, as in Figure 55, with only sufficient turning to avoid cutting pockets in the bank, so that the pit floor is easily kept smooth. The machine, with bucket on the ground, is almost twenty feet long, so trucks should be parked at least thirty feet from the point of digging. Trucks should be parallel to the bank so that loading can be done from the side as dumping range is short.

If the walking distance is short, the length of the cycle is determined by the time it takes the bucket to move from dig to dump and back, as the tractor will have to be stopped, usually after it has reached the truck, for completion of the dump; and again to await return of the bucket to digging position.

If the distance is long, the bucket is checked at some point where it will tend to balance the tractor while it walks to the dumping or digging point. This reduces stresses on the front rollers. The distance which the bucket must be carried while stationary is waste motion and slows the cycle.

This type of machine is ordinarily backed until contact is made with the truck to be loaded. The operator cannot see into the truck unless the sides are quite low, but the loader can dump at heights up to nine feet.

These units are intended primarily for loading dirt, gravel, or clay from banks, broken pavement from roads and streets, and other heavy digging from a relatively flat surface. Under average conditions, they have a larger output than front loaders of similar size, and do not subject the tractor to twisting strains.

They are well suited to excavation in driveways, alleys, or other narrow spaces where side clearance is limited. They require a free working height of about eight-

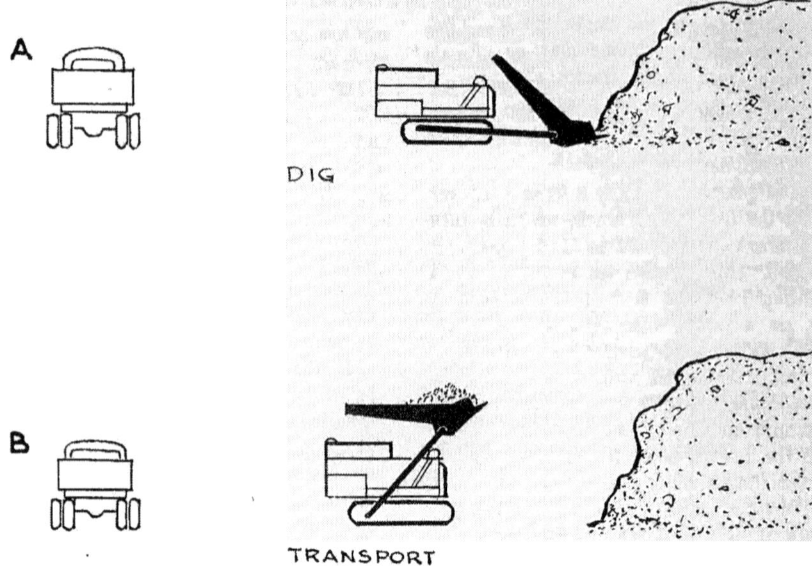

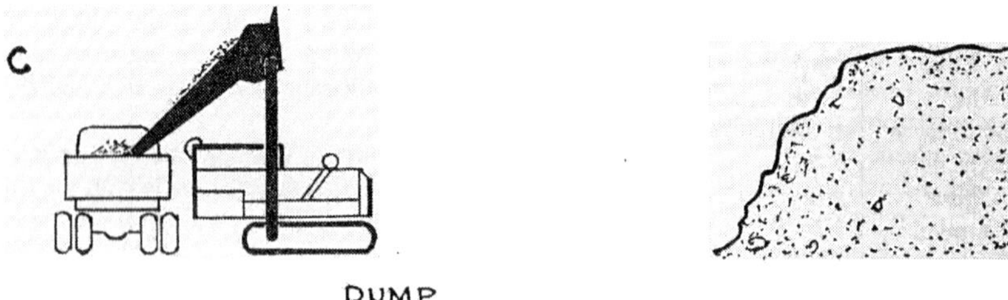

Fig. 55. Overhead loading

een feet for dumping, but can dig wherever there is enough headroom for the roof, and walk to a clear space to hoist and dump.

An "overhead" is not as versatile as a front end loader. When equipped with a bucket, it must go through the whole overhead cycle in order to dump. It can therefore grade only by using the filled bucket as a dozer, or by turning in a half circle and backing to dump the load ahead. Sometimes it is possible to keep the cutting sufficiently ahead of the filling to enable the shovel to dig to the front and dump in holes behind it without turning.

A standard type bulldozer blade can be installed on the push arms instead of the bucket.

CHAPTER FIVE

HAULING UNITS

BOTTOM-DUMP SCRAPERS

Scrapers, known also as carrying scrapers, pans, or cans, and by various trade names, are digging and grading machines which carry or drag their loads behind the prime mover. They have a very wide range of types and sizes, and include one quarter yard hand scrapers, and self-powered machines carrying thirty yards or more. They may be towed behind crawler or wheel tractors, or drive themselves; may have two axles, or one, or none; may dump underneath or to the rear, and use cable, hydraulic, or hand controls.

The bottom dump two axle machines are of primary importance in earthmoving. They are the standard tool for alternating cuts and fills, under the wide range of conditions where soil is firm enough to support them, and is fine and soft enough for them to handle or can be made so by rooters or explosives.

The scraper can dig, haul, and spread in a single normal working cycle. It can work alone if necessary, but production is usually increased if it is assisted by other machines. It works in thin layers both in the cut and on the dump, without limit as to the number of layers, so that its efficiency is not particularly affected by depth of cut or height of fill. Its use causes considerable compaction of fills, and favors proper use of rollers.

In various models, its economical haul distance ranges from less than two hundred feet to a mile or more. Where conditions are favorable, it can move earth at lower cost per cubic yard than any other type of earthmover now available.

It is not only an excellent machine for bulk earthmoving, but a precision finishing tool as well. The cutting edge is carried between front and rear wheels, so that it is unaffected by pitching, and the operator can control its position very accurately. If job conditions give him time, he can cut or fill accurately to grade, and when space is wide enough for maneuvering, can build crowns and slopes as well.

There are two basic types in use. Figure 56 is a full trailer model, which is towed and controlled by a separate tractor, usually of the crawler variety. Figure 57 shows a self-powered model, in which the prime mover is a wheel tractor more or less permanently fastened to the scraper. This tractor may be the two wheel type shown, or have a pair of front wheels also.

Crawler-powered scrapers have maximum digging power. Wheel-powered units usually need help from a pusher tractor to get full loads, but are much faster on the haul and can therefore be profitably used on longer runs.

The scraper has three basic operating parts. The bowl is the load-carrying part of the body, which is equipped with a cutting edge on the front of the bottom, and can be raised and lowered. The apron is the front wall of the bowl, and can be raised

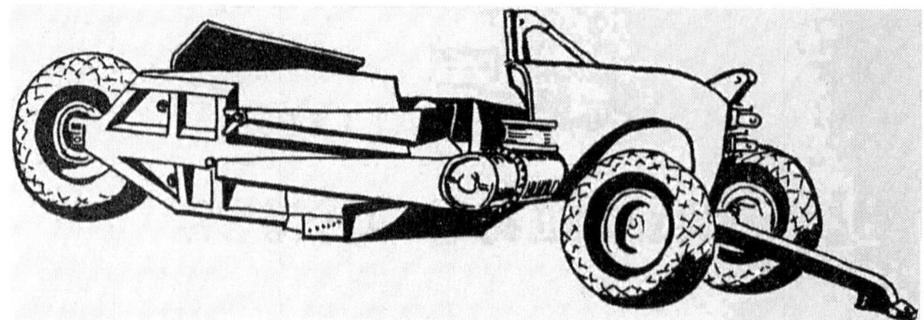

Fig. 56. Trailer scraper

and lowered independently of it. The tailgate or ejector may be the rear wall of the bowl, which is moved back to make room for a load, and forward to discharge it. In other models, it is the floor and rear wall of the bowl, and is tilted upward and forward to dump.

Control is usually by means of two winches mounted on the tractor, but three winches on either the tractor or the scraper may be used. Hydraulic controls are also available for small and medium size machines. In two-winch and in some hydraulic models the apron and tailgate are operated by a single lever through linkage arranged so that the apron opens before the tailgate moves forward, and the tailgate moves back before the apron closes.

Connecting. Two-axle cable scrapers (sometimes called four-axle or full trailer) are usually stored with the bowl blocked up and the cables in place. To reconnect, the tractor is backed to within a foot or two of the tongue and the cables are anchored to the power control unit drums.

If the tongue is light or counterbalanced it can be raised by hand and the tractor backed into it. If it is too heavy to lift, a pulley or snubber should be fastened to one side, and one of the cables looped under this projection. Tightening the cable will then raise the tongue. The tongue may also be lifted with a jack or pry.

The tractor drawbar pin should be locked in place with a bolt or cotter pin.

Hydraulic two-axle scrapers depend on counterbalancing or jacking the tongue, or leaving it blocked at correct height when disconnecting. One-axle hydraulics allow manipulation of hitch height by applying down pressure to the bowl.

Cable operated one-axle scrapers should be carefully blocked at the correct height when they are disconnected.

Loading. The tailgate is fully retracted, the apron partly open, and the bowl lowered until the cutting edge rests on the ground. The tractor is placed in low gear, or sometimes in second, and driven at full throttle. The depth of cut is regulated by raising or lowering the bowl so that the edge will cut a good slice of ground.

Except when working near final grade, the cut should be as deep as the tractor can handle without lugging down badly or spinning tracks. A deep cut fills the bowl fast, and the thick slice is effective at heaping it.

The dirt dug piles up in the bowl, some of it falling forward on the apron. Incoming dirt must be forced to rise through an increasing depth in the bowl, and a point of refusal may be reached where the tractor power is not sufficient to force any more dirt into the bowl, or when the soil will not take the thrust, so that it breaks up and is pushed ahead, or drifts off the sides.

If a full load has not been obtained, some may be added by pumping. The edge is raised sufficiently to decrease the draft

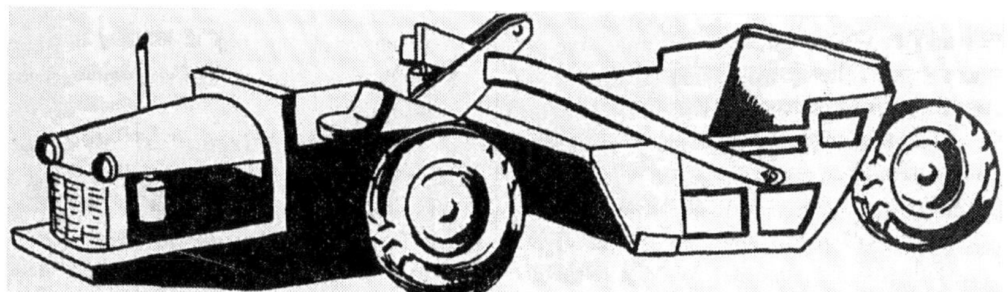

Fig. 57. Self-powered scraper

and allow the tractor to regain its full speed, then dropped several inches below the normal cutting depth. Some soil will be forced into the bowl. The knife is lifted as the tractor starts to labor and the process is repeated if desired.

The advantage obtained is partly the momentum of the lightly loaded tractor and partly the ramrod effect of the thicker layer of dirt punching its way up through the load.

This punching effect is important. It will be found that clay and other heavy soils may be loaded most effectively in thin layers which reduce the cutting power necessary, without sacrificing too much thrust. Sand, however, must be taken in deeper cuts and requires pumping with a much smaller load. It is sometimes helpful to set the apron to drag on the surface of loose soil to compact it while loading.

Scrapers are efficient at penetrating fairly hard soils, as the cutting edge is sharp and held at an effective angle. The machine cannot be overbalanced by suction, as the knife is carried between the axles, but a plastic soil may pull the edge a few inches deeper than intended by flexing the tires or causing them to sink into the ground.

There are hardpans and rocky soils which the knife will not cut and many others which can be dug only by the expenditure of so much power in penetrating that little force is left for the loading. In such cases, the use of ripping machines or explosives ahead of the pans is advisable. Scarifying costs can often be reduced and loading benefited by using widely spaced teeth, so that the ground will be coarsely broken rather than pulverized.

In hard digging, a straddle loading sequence is often helpful. Parallel cuts are made, leaving a ridge between which is narrower than the bowl. The ridge is then taken out on a third pass, and it will be found that the digging resistance is more nearly in proportion to the shallow cuts at each side than to the deep one under the ridge.

This method should be used with caution near the edges of cuts as it may destroy the crown.

The outer edges of a cut and the bank slope should be determined before starting work so that steep banks may be cut to final grade from the first. They are taken down in a series of steps. If the slope is one on three, and the pans are taking six inch slices each new cut should be eighteen inches further from the bank than the previous one.

The slope should be checked frequently by engineers for correctness, and trimmed off by a grader working on the floor of the cut, as it may become very difficult to reshape when the floor has been cut too far down.

The floor of the cut should slope down toward the edges. This slope may be originally established by an angle dozer or a grader, after which it will tend to perpetuate itself, as the weight of the machine will be greater on the down side so that it will tend to cut low there. If the slope becomes too great the upper part may be readily planed off.

Machinery may not be available to shape the original surface, or the crown may be lost because of oppositely sloping strata, or by careless operation. A scraper can cut a crown by taking advantage of the fact that the oscillating front axle does not affect the side-to-side tilt of the knife, which is determined almost entirely by the rear axle. A gouge taken heading up a slope can be used to tilt the rear axle so that the knife can cut on the uphill side when turned along the slope.

Whenever possible, the cut should be arranged for digging downhill and toward the fill. The first factor is usually more important. The grade of the cut is most important when it is or can be made steep, and when power and traction are small in proportion to the size load desired.

Pushers. Except when using a tractor which is oversize for its scraper, loading down steep slopes or in exceptional material, pushing a scraper with another tractor greatly increases production by speeding the loading and increasing the size of the load.

Most pushers are crawler tractors, although there is increasing use of rubber-tired four-wheel drive machines. Because of their greater speed, these can service a larger number of scrapers under average conditions. However, they ordinarily do not exert as heavy a push and their usefulness declines rapidly if the cut becomes slippery.

Scrapers can also load each other. Transporting a scraper, whether full or empty, takes only a fraction of the power required to pick up a load. A prime mover with a scraper attached can therefore give a powerful boost to another unit either before or after it picks up its own load.

A bumper may be put on the back of a pusher so that it can be pushed while pushing.

The pusher tractor is often dozer equipped, and can do rough grading and odd jobs between scrapers. It is good practice to reinforce the center of the blade, either internally or with a pusher cup on the surface to prevent caving in the moldboard with repeated maximum strain and shock in the center.

There are also special pusher plates which are rigidly attached to the frame of the tractor, and which cannot be moved up or down except by hand adjustments. These are much cheaper, but may cause difficulty in lining up with bumpers on different makes of scrapers, and may be difficult to keep in contact during pitching caused by very rough ground or changes in torque.

The pusher is driven up behind the scraper, and contact made with its bumper as smoothly as possible. The scraper unit applies all its power, while the pusher uses as much as it can without causing the scraper to jackknife, as it will if the pusher and scraper move faster than the scraper's tractor. Jackknifing must be guarded against, as it will often damage the scraper tires by contact with the blade or tracks of the pusher.

Transporting. In transporting loads the bowl should be high enough to avoid collision with the ground or objects on it. However, it should never be held against its stops. This may damage sheaves by pulling them into each other, and it leaves no give in the line so that when stretched by a sharp turn or by weaving of the frame on rough ground it will be likely to break.

Stability on turns and side slopes is increased by keeping the bowl low. On extreme slopes it is even allowed to drag.

On graded haul roads the highest gear that will move the load without lugging down is used, and on rough going the highest which will give enough stability and control.

Although both track and wheel tractors can drag loaded scrapers successfully over soft and rough ground, they do it at the expense of speed and maintenance expense. Yardage, both in the load and in the number of loads can be greatly increased by providing a smooth, hard surface for rubber-tired tractors, and a smooth, slightly soft surface for track types. If a very large quantity is to be hauled on rubber it should pay to oil or blacktop the road.

For short period use by rubber, or any use by crawlers, the road top may be dirt unless wet weather is expected, in which case bank gravel, shell, or other material not seriously softened by water should be used. The road should be maintained by a grader, but if none is available it can be kept in shape at somewhat higher cost by the pans. In dry weather watering equipment will be required.

Upgrades slow the movement of loaded pans, the amount of loss depending largely on the ratio of the power of the tractor to the weight of the load. Self-loading rigs will get over the grades faster than pusher loaded ones of the same type, and on severe grades level loads may make enough better time than heaped loads to give higher production.

Soft footing absorbs a lot of power and may make a critical difference in uphill hauling, particularly to wheel tractors.

Unless footing is very bad rubber-tired units will make much better time on the haul.

Steep down grades slow scrapers, as they should be descended with caution, particularly when carrying a load. Full trailer scrapers generally have no brakes and may jackknife if the tractor stops or slows down sharply.

Sharp turns should be made slowly, as the momentum of the scraper tends to carry it in a straight line, and if the tractor is not in that line and speed is high, a complicated and sometimes serious wreck may follow.

Self-powered scrapers with front brakes only may jackknife when stopping quickly, or descending steep grades against the engine. If the bowl is dropped, it will both slow and straighten out the machine. Those with rear brakes only may get out of control if allowed to roll down a grade backward.

Dumping and Spreading. To dump, the bowl is lowered until the edge will just allow a layer of dirt of the desired thickness to slide under it. The apron is then raised enough so that the falling dirt will supply the knife sufficiently to make a continuous smooth layer, without dropping excess which would heap in front, or drift to the side to make windrows.

When the apron is fully raised the tailgate is moved forward gradually, pushing or dumping the dirt out of the bowl. Too fast a crowd will supply too much dirt to the knife and will also place unnecessary strain on the cables and sheaves of the ejector mechanism. If the dirt is sticky good results are obtained by advancing the gate a foot, allowing it to slide back six inches, and advancing it again when required. A lift gate may be banged against its stops to knock it clean. Dirt resting on the edge after the gate is fully forward may often be dumped by moving the gate back and forth.

When the dump is complete the gate is fully retracted, the apron dropped, and the bowl raised to carrying position.

Spreading is usually done in thin layers as it provides better compaction and eliminates or reduces the need for other grading equipment. It also keeps a more even grade on the dump, reducing the hazards of high speed work.

Dumping and spreading is ordinarily done heading away from the cut and at medium or high speed. In finishing a grade and under some other conditions a low gear may be used to give more exact control.

The fill should be started at its outer edges and kept built out to finish slope all the way up. This is important, as the type of pan ordinarily used has no way of dumping over edges and any patch fill made could not be readily compacted, so as to support the machinery building additional layers.

It is also desirable to have the fill slope up toward the edge, as this arrangement will place the larger part of the weight of both tractor and scraper on the side away from the slope, and reduce danger of caving. However, if rain is expected, the fill may be crowned enough to shed water.

The closeness with which a pan can approach a high edge is determined by the type of soil, the degree of compaction, the slopes, the weight of the machines, and to some extent by the skill of the operator. In general, heavy soils which are not in a muddy condition are safer than sandy soils. Working right to the edge or overhanging it slightly with a tamping roller will give good compaction.

If conditions are such that the scrapers cannot safely go near enough to the edge to build a proper slope, they can drop the material a few feet in and a grader can distribute it.

The slope should be checked frequently for proper grade.

Side Hills. If a cut slope is moderate it can be smoothed off by pans, dozers, or graders, either in strips or after completion. Such a grading unit may be used on steeper slopes if supported by chains to a machine moving parallel to it on the top.

A full trailer pan should not be used on steep side slopes unless the operator is very skillful, or other equipment capable of righting it is on the job. The narrow front tread, oscillating front axle, and tendency to jackknife when backed up all make it liable to side tipping. It is also unsafe to use it close to the edge of poorly compacted fills.

If side hill work cannot be avoided, a scraper with minimum overhead structure should be used and the bowl should be carried low. Turns should be made uphill where possible and downhill turns should be made gradually, preferably with the apron fully down and the bowl scraping on the ground.

Danger of tipping is particularly severe if the soil under the upper side of the machine is harder than that under the downhill wheels or if the upper side is forced up by running over boulders.

Overturning a trailing scraper usually does little or no damage except in lost time, as the swivel in the draw tongue allows it to turn freely and the structure is sturdy enough to resist damage. A good operator, or a lucky one, may be able to right it by towing it over a hump or the edge of a hollow that will roll it back. If the cables are not too tangled the parts may sometimes be worked into a more suitable position before towing.

If it cannot be righted by towing it can be rolled by pulling a high part from the side, preferably by another machine. If none of sufficient power is around, the tractor must be disconnected in order to get beside it to right it.

Work Patterns. Use of an efficient work pattern is important in obtaining maximum production. Such patterns are sometimes dictated by the job layout with little opportunity for change, but at other times the work can be done in several different ways.

If a mixed lot of machinery is available, it is helpful to assign each unit to a length of haul and type of work suitable to it. In general the dozer does short-push work, the crawler-drawn scraper the medium hauls, and the rubber-drawn scraper the long trips. These statements are for average conditions and will not hold for all jobs, as using oversize pans, or necessity of traveling on very soft footing and up steep grades give the crawler rigs an advantage even on fairly long hauls, while ideal loading and haul conditions will give wheel tractors an edge.

However, great savings may be effected if the fast machines on rubber can be kept on long hauls, the crawler ones on short hauls, and the dozers kept to trail breaking, reduction of very short rises, and cleanup work. This, however, will involve continuous operation of at least two cuts and one fill, or one cut and two fills, unless a fill is long enough so that the long range machines can do the far half, and the crawlers the near section.

If two cuts are available, but only one pusher, the crawlers can self-load at the nearest cut to the fill and the pusher can help the wheel rigs at the further cut.

In many cases the savings of such division would be outweighed by the extra supervision and traffic difficulties involved.

Savings may be effected by reducing the time loss in turns. A scraper of any kind may take from ten to sixty seconds to make a turn-around. The fastest turns are made without a load when the space is wide enough for an open swing, and flat enough so that tipping over is not a problem.

Another time loss in turns is the distance which must be traveled beyond the completion of the dump or past the loading spot to reach the turn. This loss is often increased unnecessarily by careless dumping or loading which leave the fill or cut too rough for a turn, or by placing too many grade stakes to permit turning in the work area.

Even if the turn is efficiently handled it is a dead spot, and on a short haul may take a substantial part of the cycle time. When a succession of cuts and fills are to be made, it may be possible to increase production by reducing the number of turns.

Turn patterns are of course subject to the overall plan for the job which may include some very complex factors of distribution, work sequence, and separating of different types of soil. However, an alert supervisor can usually effect an increase in production by seeing to it that turns are kept to a minimum, and that adequate space as close to the work as possible is provided for those which must be made.

When a pan loads it pushes some dirt in front of it, more in loose or sandy soils than in heavy ones. This pushed heap may be used to build up fill areas which are at the edge of a cut.

Grading. Scrapers may also be used to advantage in grading and leveling areas where the quantities of dirt to be moved are not enough to fill the bowl.

The location of the knife between front and rear supports gives the pan good stability for grading, and controls are sufficiently sensitive for fine work. It cannot shift dirt from side to side or tilt the blade as a grader can, so that crowning and finishing a road is more laborious.

On the other hand, the pan can move substantial amounts of material along a road, and can bring in borrow, or dispose of surplus off the road. A pan and a grader working together are a very efficient team for rebuilding haul or country roads.

When cuts and fills are shallow and closely spaced, the usual working position is with the apron fully lifted, and the tailgage almost fully forward. It is bad practice to have it all the way forward, as it will not leave any give in the apron and tailgate cable, so that parts may be strained or the cable broken on turns or rough ground.

The bowl is lowered or raised to give the desired depth of cut or thickness of fill, and acts much like a dozer or grader blade.

If a cut is deep or long enough to provide enough spoil so that it starts to drift off the sides, the tailgate can be retracted or lowered enough to admit it inside the bowl, and then used to push it out again when a fill is reached.

There is no definite line of separation between light grading and heavy cuts and fills.

Breaking Land. The scraper is often dependent on other equipment, particularly dozers and rooters, to prepare an area so that it can work in it. Ordinarily, the towing tractor is not dozer equipped or the dozer is useless because the control winches are used for the scraper. If the dozer is usable, a crawler is still greatly handicapped by the attached trailer as it cannot be backed more than a few feet without the most delicate manipulation.

Wheel tractors with the semi-trailer type of scraper can maneuver freely but do not have the traction for heavy dozer work.

Lacking a dozer, the tractor is compelled to walk on the original grade which may be dangerously steep, rough, or overgrown, and try to manipulate the scraper behind it at the same time.

When a dozer is clearing for scrapers it should remove all trees, heavy bushes, stumps, and boulders. Gullies should be broken down and smoothed in. Hillcrests should be rounded off or flattened so that hauling equipment can cross them or turn around on them. Side slopes should be notched at the upper edge of the cut to produce a slight slope opposite to the original.

If there is no access to the top of a ridge even for dozers, it may be cut off by a dipper shovel working from the highest convenient access. The shovel may cut to grade or it may simply provide a road which dozers and then pans can use and deepen. In either case it should cut to full width.

Normally trucks or wagons are used to haul from the shovel, but pans which do not have overhead structure to interfere can be used. Shovel loaded dirt is much looser than that scooped up by the bowl, so the bank yards carried from the shovel will probably be fewer and the load lighter than if the pan dug it itself, even if it is heaped as high.

Stumps and Boulders. Scrapers are not designed for moving of boulders and stumps which can usually be handled more economically by dozers or rooters. Some makes will stand up well under this type of work where others may give trouble with bending and breaking of parts.

If the stump or stone is low enough the tractor can walk over it, drop the bowl with the edge two or three feet behind it, and move forward. If the edge hooks into the object, it is lifted and forward motion continued. If it slips off the tractor is stopped immediately and backed for another try.

If the blade hooks in but will not move it, the steering clutches are released, the bowl raised to break it out, and the clutches re-engaged.

The apron is carried at full height to keep it from hitting. If the object is to be picked up and carried the gate is kept full back, but if it is simply to be rolled out the gate is kept far enough forward to prevent it from entering the bowl.

Often the stump or boulder cannot be picked up on the same pass that loosened it, so that the pan must be backed and another bite taken. It is quite difficult to back a full trailer scraper, so the forward movements must be kept to a minimum until the rock is well hooked.

If the machine cannot be backed into position for another bite, it must be driven in a circle to come up from behind. Often

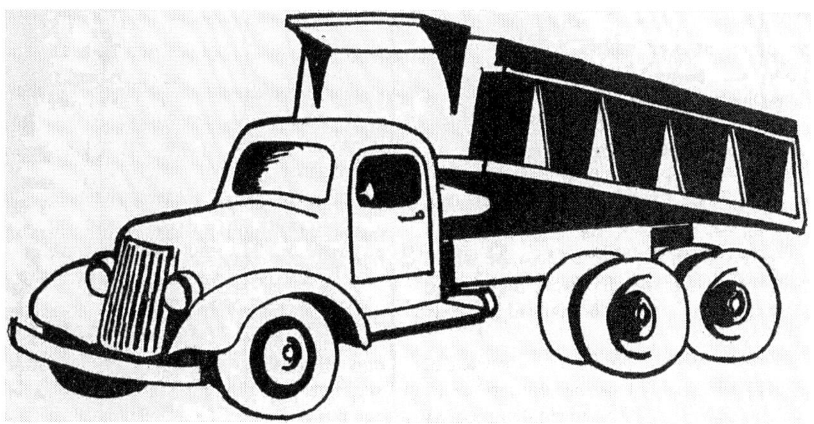

Fig. 58. Dump truck

it will not be possible to drive over it again, as the work may cause the obstruction to turn into a higher position.

In that case, or if the original position was too high for the tractor to go over, the tractor is driven past the object, close to it. It is then cut in so that the front wheel just grazes the stump, then turned further so that the edge is brought into position to work on it.

Scrapers, particularly of those models which have very high apron lifts, can pick up large stones or stumps. However, these sometimes turn or catch when in the bowl and become jammed when an effort is made to discharge them. They can be worked out by moving the gate back and forth and the bowl up and down; by digging enough dirt to shift its position; by pulling the machine onto a favorable slope, or overturning it; or prying with bars or logs, or reducing the object with axes, chisels, sledges, or rock drills.

Oversize rock is liable to cause damage by denting, bending, or straining parts. Damage may also be done by accidental collision with boulders during ordinary digging or transporting.

DRIVING A DUMP TRUCK. Driving a dump truck is somewhat like driving a car. However, the truck is much heavier and more bulky, has very poor visibility to the rear, and usually requires special methods of shifting gears and climbing and descending hills.

In addition, checking the truck, spotting it to load and dump, and operation of the body are special techniques not related to any experience gained driving a car.

Checking. It is particularly important that tire inflation and condition be checked at least daily, as the tires are the most vulnerable and most expensive wearing parts of a truck. Running them soft with a load may ruin them in a few minutes.

If the truck has air brakes, it should not be moved until the dash gauge shows safe operating pressure.

Direction signals and stop lights must be in working condition for driving on highways or on heavily traveled haul roads.

Stones stuck between dual tires should be removed immediately as they may be thrown behind the truck with deadly effect. A stone can usually be pried out with a bar or board, but it is occasionally necessary to let the air out of one tire to free it.

Each load should be checked for rocks

or other objects which might fall off. These may be moved to safer positions or be pushed off in the pit.

Loading. Truck drivers may add substantially to shovel production by getting promptly in proper position. If a spotter is present or the shovel operator is giving signals, directions should be followed exactly regardless of the driver's ideas of where the truck should be. If a spot log is used the rear wheels should be backed squarely against it.

If the shovel has not moved a truck will generally be spotted in the same place as the last one loaded. Following tire tracks or observing the location of spilled material makes it easy to get in the same position.

In general, a truck hauling from a revolving shovel is placed to require the shortest practical swing from the digging. It should be facing directly away from the shovel so that the shovel can reach from back to front of the body by use of the crowd mechanism, or at a right angle so that the swing controls can be used to distribute the spoil along the length of the body.

When large dipper shovels are loading rock the truck is sometimes placed facing the shovel. The open door of the bucket tends to prevent rock from rolling toward the shovel and therefore serves to protect the cab and the operator.

The distance from a dipper shovel should be such that the center of the body is in the middle of the crowd arc of the bucket. With draglines and clamshells it should be directly under the boom point. A hoe loads fastest if the truck is on lower ground or very close in.

When being filled by a front end loader, the truck should usually be backed against the bank at an angle of forty-five degrees so that the tractor need make only one eighth of a turn to dump. The truck should be close and the loader walking time should not be greater than the time required to lift the bucket. Wheel tractors require more space to maneuver than crawlers.

For overhead loaders, the truck should be parallel to the bank and far enough away to allow the machine room to maneuver but not so far that it must stop the bucket hoist while walking.

It is often necessary to move a truck while it is being filled. This is particularly true with belt or bucket loaders, which may themselves be moving, or if stationary, may not distribute the load throughout the body.

The driver can remain in the cab during loading, but a broken shovel cable may allow the truck to be struck by a falling boom or an uncontrolled bucket. The size of the shovel and the amount of cab protection largely determine the amount of danger. A backhoe loading from behind the truck is particularly hazardous.

In pulling away from the shovel the clutch should be released gradually in low gear. The throttle should be opened enough to avoid stalling, but racing of the engine and spinning of wheels in soft pits should be avoided.

Gear Shift. The majority of trucks now in service have sliding spur gears in the transmission. In order to avoid clashing when shifting it is necessary that the two gears to be meshed have about the same tooth speed. Shifting arrangements are quite variable. Figure 59 shows six patterns which are widely used.

When the truck is standing, the clutch is pushed down and after a slight pause the gear lever is moved to the first or low gear position. The pause gives the jackshaft, extending from the clutch into the transmission, a chance to slow down or stop so that it can be meshed.

After the truck is moving, the shift to second is started by releasing the accelerator, depressing the clutch, and moving the

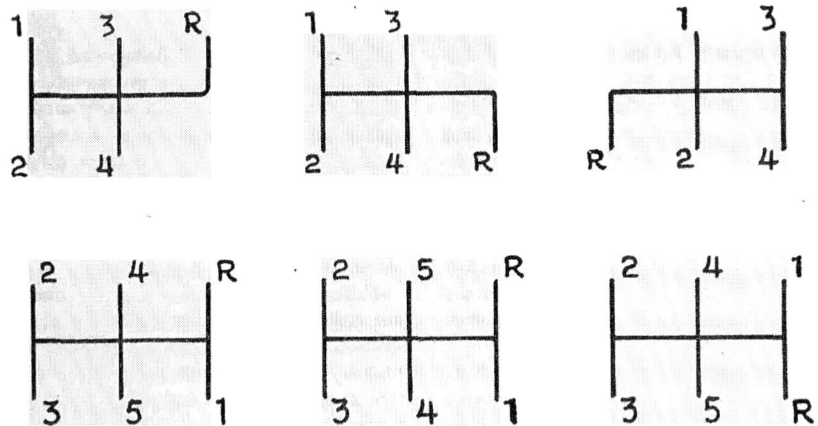

Fig. 59. Gear shifts

gear lever into neutral. The teeth on the sliding gear will be moving at the same speed as those on the countershaft first gear, but more rapidly than those on the smaller countershaft second gear. If the clutch is held down the jackshaft will slow until its sliding gear can be meshed quietly into second. However, the engine slows down much more rapidly than the free-spinning jackshaft, so that if the clutch is engaged again after the shift to neutral and then depressed, the jackshaft will be slowed so that the shift can be completed immediately upon releasing the clutch the second time.

This operation is known as double-clutching and is used with any shift to a higher gear. It does not work if the engine does not throttle back to idling speed.

If a truck has a properly adjusted clutch brake it will slow the jackshaft without the necessity of double-clutching.

When a shift is made from a higher to a lower gear, as in climbing a hill, double clutching is required even if there is a clutch brake. In this case the jackshaft gear will be turning too slowly to engage with the next lower gear, therefore the clutch is pushed down, the gears put in neutral, the engine speeded up, the clutch let up and pushed down, and the shift completed.

Shifting up is mostly a matter of timing, but shifting down requires both correct engine speed and timing and is somewhat more difficult to master. It should usually be done quickly before valuable momentum is lost by the truck.

If the truck has an auxiliary transmission it is usually of the two speed, sliding gear type. Shifting is done in the same manner as in the main transmission. However, the jackshaft has much more inertia because the main jackshaft and transmission gears are spinning with it, so if a bad shift is made, the danger of damage to the teeth is greater.

If a transmission is of the synchromesh type, double clutching is not necessary unless the synchronizing mechanism becomes ineffective.

Axle Shift. Many trucks are equipped with two-speed axles with power shifting. These are controlled by a valve or switch on the dashboard or the gearshift lever. The power may be vacuum, air, or electricity.

These shifts are of the pre-selective type. A truck in low axle ratio may have the control moved to high axle, but as long as the engine is driving the truck the pressure on the teeth will prevent the gears from disengaging. If the accelerator is released, the tension will be relieved, and the power mechanism will move the shifting collar or other device out of the low axle position and push it toward engagement with the high axle. There will be a discrepancy in speed, but the driveshaft loses speed rapidly, and the shift should be completed automatically when speeds are synchronized.

In the shift from high to low, the hand control is shifted, and the accelerator is released and then pushed down.

When going downhill against the engine the shift is made by pre-selecting the ratio and opening the throttle briefly to reverse the torque.

A "split shift" is shifting transmission gears and axle ratio at the same time. This is done by pre-selecting the axle ratio and depressing the clutch and shifting gears in the usual manner. The axle shift will take place while the clutch is out or immediately after re-engagement.

Hills. A truck with a gross weight of 16,000 pounds may use the same engine as a car weighing 3,500. Brakes, although larger, are not increased in size in proportion to the extra weight which must be controlled.

Hills, therefore, present two problems to trucks—of pulling up and of getting down safely. It is sometimes possible to rush small hills and get over them in high, and to descend moderate grades in high gear with use of the brakes.

Generally, however, the gear shift must be used. When the truck slows below the efficient torque point of the engine (usually between one third and two thirds of top speed in the gear being used), a shift should be made to the next lower gear. If the shift is delayed too long the next lower may not be able to pull so that another shift down must be made immediately.

If the truck is overloaded or the grade very steep the truck may not be able to start moving from a dead stop, so that unless gears are shifted properly it may become necessary to back the truck to the bottom again or to get help. Under such circumstances the amateur truck driver may do well to put it in low before he starts up.

Failure to shift properly going downhill may have much more serious consequences. If too high a gear is used the brakes may become overworked so that the drums heat and expand away from them. If the brakes are loose, the pedal may strike the floor or require "pumping," before it can make an effective contact. If they are tight, there is still the danger that the curve of the shoes will no longer fit the expanded drums, so that not enough lining will be in contact to stop the truck. Heat glaze may destroy the effectiveness of the lining.

If a shift down is attempted when the truck is moving too fast, it will be difficult to complete both because of the speed and its mental effect, so the gears may remain in neutral. A truck going downhill out of gear is even more out of control than when in high gear.

Proper procedure is to slow to a near stop before starting down a grade and to shift into a gear low enough to hold the truck to a safe speed with only occasional braking. If the speed still increases too much, a near or full stop should be made before a shift into a lower gear.

Downhill speed should be kept within the normal range of the gear being used. If the engine is allowed to "wind up" past its maximum operating speed severe damage may result. If the truck is allowed to coast rapidly in a low gear with the clutch out, the clutch may be wrecked by centrifugal force.

Hill problems are directly related to load. An empty truck needs only a little more care than a passenger car, while a severely overloaded one may be unsafe even with the most careful handling.

Curves. Curves must be taken more slowly with a load than without. The truck should be slowed before entering the curve and accelerated or held at an even speed while in it.

When rounding a curve the centrifugal force tends to pull the truck toward the outside, and this force is resisted by the friction of the tires on the road. If traction is good the truck will tilt, compressing the outside springs and tires. Increase in load will both raise the center of gravity and increase the force tending to tip or slide the truck.

The force required to turn the truck is exerted by the front wheels, which slow the truck and cause it to nose down very slightly. The outside wheel does more than half of the turning work because the tip of the truck adds to the weight it is carrying. This wheel is therefore under the greatest strain of any part of the truck.

When brakes are applied, the forward momentum of the truck is resisted by the wheels. This transfers weight forward, compressing the front springs and relieving the rears. This increases the burden of the front wheels, particularly the outside one which may be already overloaded.

Compression of the outside front spring by both turning and braking forces throws the truck off balance and makes it difficult to control. The overloaded wheel may roll its tire, hop, or skid.

Accelerating the engine causes the rear wheels to push forward against the inertia of the truck, which tends to make the truck nose up, transferring some weight from the front wheels to the rear ones, and improving stability.

If the road is banked, the centrifugal force acts to pull the truck to the pavement so that stability is improved. Reverse banking, as in a left curve on a crowned road, will cause it to pull the truck away from the road.

Skids. In normal operation the direction in which a wheel and tire are rolling controls the course of the vehicle and considerable resistance exists against forces tending to slide the wheel sideward. This resistance decreases if the wheel is turning either faster or slower than required by the speed of the truck. If the wheel is locked it moves sideward almost as readily as along its plane of rotation.

If the front wheels are locked while rounding a curve they will no longer turn the truck which then moves straight ahead. This type of skid most often occurs when brakes are applied too hard, but on very slippery surfaces the wheels may lock when turned sharply without brakes. When the brake is released the wheel may revolve again, or may remain locked until turned in the direction of the skid so that its friction with the road tends to revolve rather than bend it.

When the front wheels are again revolving they will substantially control the direction of the movement of the front of the truck until again locked by brakes or too sharp a turn.

If the rear wheels lock on a curve and the front ones do not, centrifugal force causes the rear of the truck to slide outward, pivoting on the front axle. The brakes should be immediately released and the front wheels straightened so that the side pull of the turn will be stopped.

If the rear wheels have skidded too far sideward to be started rolling in this manner, the tendency of the truck to spin may be checked by steering sharply in the direction of skid. The front wheels will travel in the same direction as the back is sliding, and the truck will move diagonally sideward. The front wheels will tend to get in the line of skid, and the rear wheels to start turning. If the engine stalls it should be started immediately.

Should the rear wheels slide due to spinning, traction can usually be restored by releasing the accelerator and pressing it down gradually.

Sometimes it is more important to avoid collision with a person or a particular object than to straighten out the skid. Also, once a truck has started to "pinwheel," straightening it out is a matter of skill and luck rather than rules. The safest system is to avoid skidding by driving slowly and cautiously when roads are slippery.

Backing. Most fills using rear dump trucks require turning and backing to the edge of the fill. If possible, a fill should be arranged so that the turn is made near the dump spot, the turn spot should be wide enough so that reverse gear need be used only once, turns in reverse should be toward the driver's left, and the truck should be level or facing uphill while dumping.

If the truck has a two speed axle or an auxiliary transmission, two reverse speeds are available. Low should be used when the load is heavy, the ground soft or rough, or when complicated steering is required.

Vision to the rear is blocked off by the body. The cab guard may be solid so as to render the rear window of the cab useless. If there is no guard, or it has eyeholes, view of the ground will still be blocked by the load or the tailgate.

A view to the left can be obtained by watching the mirror, or better by leaning out of the cab past the side of the body. Unless the body is very narrow or the cab very wide, it may be necessary to open the door and stand on the left of the running board. The gas and brake are controlled by the right foot and the steering done by both hands or the right hand alone. To stop, the driver swings back onto the seat so that he can use both feet for clutch and brake.

The driver can see to the rear of the right side of the truck by partly raising the body, so that he can see between it and the frame.

Under many circumstances, such a view to the right is not sufficient for accuracy. The route must then be plotted ahead of time either before or while backing, so that steering can be done in reference to the left wheels with enough space left to keep the right ones out of trouble.

Whenever a truck (or a car) is turned while backing, the front and rear wheels will go in opposite directions from the centerline of the machine. It follows that a truck cannot get around curves in a driveway unless the drive is wider than the truck. The additional space needed depends on the length of the truck, the sharpness of the curve, and the skill of the driver. In backing, the front wheels tend to go off the drive on the outside, while in forward motion the rear wheels tend to run off on the inside.

Since the front wheels are steered oppositely to the direction of turn, a vehicle cannot back away from a wall which its side touches.

Because of restricted vision to the rear and the weight of a truck, the driver must take every precaution to avoid the possibility of children, spectators, or workmen getting behind the blind side of the truck while backing.

Dumping. When dumping off the edge of a fill, the driver should back so that both rear wheels will be the same distance from it, as in Figure 60 (A), rather than at an angle to it, as in (B). If one wheel sinks in deeper than the other, it may not be possible to either dump the truck safely or to pull out with the load.

Safe distance from the edge is determined by circumstances and the judgment of the driver. If the fill is shallow or of firm material or the truck has tandem or all wheel drive, a very close approach is possible. Certain rear dump semi-trailers

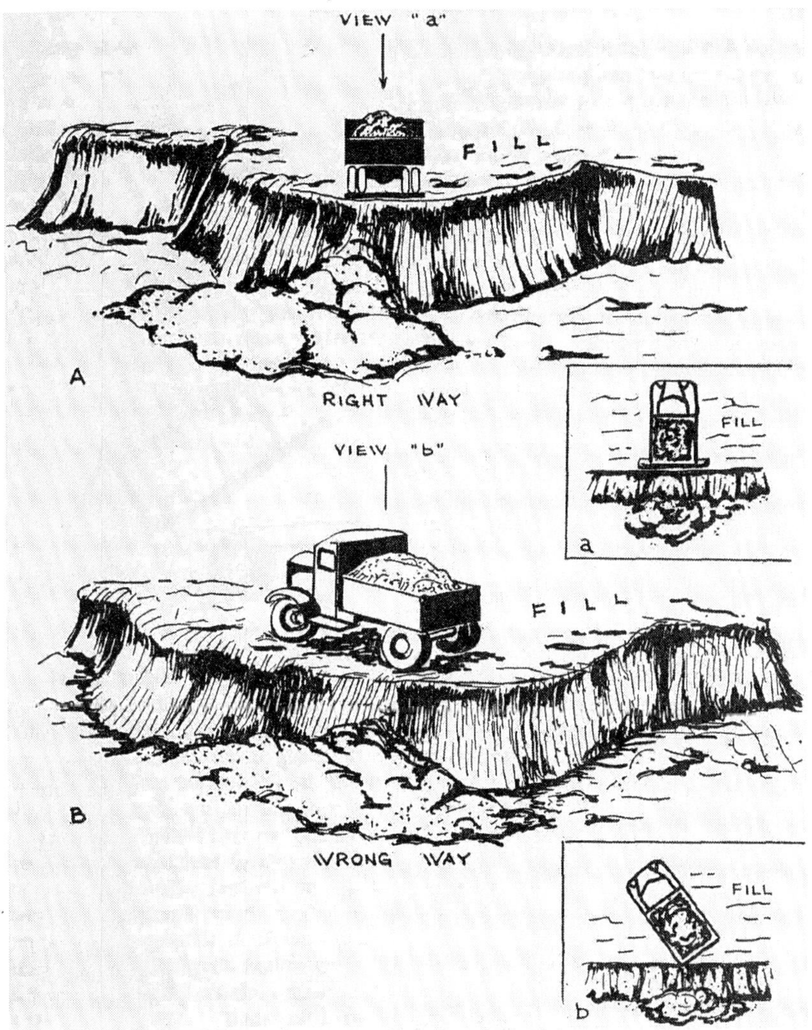

Fig. 60. At the dump

can be backed over the edge. On the other hand, it may be necessary to keep six or more feet back if the fill is soft, slippery, sandy, or otherwise treacherous. Very high fills should always be treated with respect.

Another factor is the means used to spread. If one or more large bulldozers are spreading and easily keeping ahead of the loads, the truck need not risk a close approach, but the danger is reduced as it can be pulled out readily if stuck. If spreading is done by hand or by overworked machines, dumping over the edge will speed the job but it will be less convenient to extract the truck if stuck.

When the truck is in position to dump the tailgate latch is released, the transmission power take-off engaged, with use of the clutch and a hand control, and the hoist valve placed in UP position. The engine is accelerated to a moderate speed but not raced.

Most hoists are constructed so that they can be left in the UP position to hold the body at its high position. However, if this appears to strain the mechanism the valve may be moved to an intermediate HOLD position as soon as the body is at its steepest angle.

As the body rises, the load slides backward along the floor and under the tailgate, as in Figure 61(A). Unless right at the edge of a bank the load will tend to pile up until it blocks the gate, as in (B). The truck is then placed in low and moved forward without disturbing the hoist controls until there is space for the remainder of the load to slide out.

If part of the load sticks in the body the truck may be backed into the pile to shake it loose. This may have to be done several times if the load is sticky or the dumping is uphill. The clutch should be released just before the wheels hit the pile to avoid shock to the power train. If the truck rolls away from the pile on the rebound the shaking out can be done without shifting out of reverse, otherwise low gear is used to drive forward a few feet and reverse to hit the pile.

If any considerable part of the load cannot be shaken out, it must be dislodged with a hand shovel or other tool. Most of the body can be reached from the pile and tailgate but the upper end may require climbing up on the body running boards. Sticking can be reduced if the shovel operator will put the driest material he has in the bottom of the truck so that it will slide out carrying the wetter dirt.

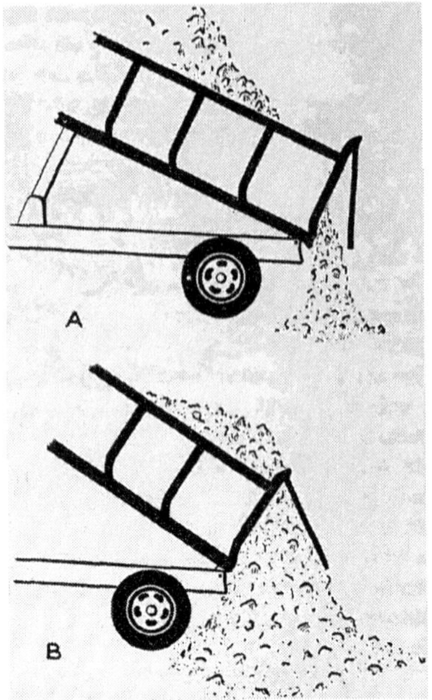

Fig. 61. Dumping a load

If the load includes rocks, stumps, or mats of trash large enough to have difficulty going under the tailgate, the gate should be left latched at the bottom and unfastened at the top so that it can drop down and the load slide over it. However, it is not safe to back against the pile to shake out the load with the gate down as it will be bent or knocked off. If much bulky material is to be carried, it is good practice to take the gate off and use higher sideboards so that a full load can be piled toward the front.

Loads put more strain on the hoist when heaped in front than when evenly distributed.

If dumping is done with the gate hinged at the top in the usual manner, and the load jams instead of sliding under it, the upper part of the body usually empties and the load piles up at the back. Conventional hoists have one-way rams which serve only to lift the body and depend on gravity to lower it. However, when the load is all in the back the body is out of balance and will remain in the dumped position until it is emptied.

The weight of dirt makes it difficult to remove the obstruction or the tailgate. Sometimes the body can be lowered by driving forward and applying the brakes sharply, with the control in LOWER. Other times it is possible to pull it down by a winch cable. Most often, however, hand tools are used to dig the dirt off the top and loosen it from the bottom until the object can be pried, cut, or smashed into a size or position which will enable it to slide through.

Trouble in dumping also comes from defects in the hoist. If it acts normally except that it will not lift to full height, the trouble is apt to be lack of oil. In this case, the body should be blocked up and oil poured into the system. A quart or two is often sufficient.

If the lift is slow and tends to creep back down if the engine is idled, the hoist control lever may have slipped. Sometimes it can be made to work properly in a different position, and at other times the linkage must be adjusted.

A loaded body should not be raised into dumping position unless the rear wheels are nearly level. Any slant will cause the center of gravity of the body to shift downhill as it is raised. The twisting force exerted may break the body hinges, tear the sub-frame off the chassis, or overturn the truck.

Spreading. The load may be spread over a considerable space instead of being dumped in a pile. The procedure is to start the dump in the manner described, and when the body is high enough to begin to spill, put the truck in gear and drive forward while continuing the hoist. If low gear is used and the hoist is fast the spread will be short and heavy. If the hoist is slow, or is so constructed that it can be slowed by opening the valve only partially, the material will be spread thinner.

Use of higher gears for the truck will also thin the spread. If the hoist cannot be set partially open and if the highest gear that will move the truck is not fast enough for the spread desired, the hoist valve can be opened and closed as the truck is driven forward.

This type of dumping is liable to result in a series of piles rather than a smooth sheet. If the load will flow freely and contains no lumps a much smoother spread can be obtained by chaining the bottom of the tailgate to the body so that it will open only a few inches.

When the body is raised so that the load starts to pour out under the gate in a sheet, the truck is driven forward. The body is hoisted just enough to keep the opening fully supplied, as raising it to full height might cause part of the load to slide over the gate or might overbalance the truck.

With any one material, the thickness of spread will be determined by the gate opening and the speed of the truck. A larger opening will increase the thickness, a higher speed decrease it.

Surface Dumping. Dumping on the surface of a fill is usually arranged so that it can be done without backing. Rear dump trucks may be used, although their height in full dump position makes it advisable to keep on smooth ground and move slowly. They may spread the load or dump in piles.

Such surface building is usually done by scrapers, or by bottom or side dump trucks or wagons.

The bottom dump wagon moves rapidly while dumping. The thickness of the spread is determined by the relation between the speed and the width of the door opening. The maximum height of the windrow is determined by the clearance of the wagon.

The wagon may be followed immediately by a dozer which spreads the material in a single pass, or a number of windrows may be built both behind and beside the front one. The latter way is economical of dozers but is liable to slow down the wagons and might cause upsets because of the necessity of traveling on the rough fill.

A side dump may be used on the surface of the fill in the same manner as a bottom dump wagon, or dump off an edge which is long enough to permit it to get broadside to it.

Trailer Operation. Driving a semi-trailer resembles ordinary truck driving in a general way, but special precautions must be taken in steering and stopping, and particularly in backing.

Very sharp turns may be made in proportion to the length of the whole unit, particularly when the tractor is a short coupled, off-the-road type, and the trailer yoke is high enough to allow space for the tractor rear tires under it. However, turns must be made carefully to avoid overrunning and jackknifing.

In a solid truck only the front wheels are turned and the momentum of the truck pushes against them at their center of revolution. The light weight and fairly positive control of the wheels, impossibility of turning the wheels beyond their stops, and the low line of push make this a rather safe and stable arrangement.

A sharp fast turn with a semi is dangerous. When the tractor is turned the weight of the trailer pushing straight ahead tends to skid the driving wheels to the outside of the turn, increasing its sharpness, so that the unit may jackknife with the two parts, still connected, trying to move in different directions. This may do no more than cause a tangle that will be difficult to straighten without a tow, but it can result in serious damage, overturning both units and tearing the hitch apart.

Jackknifing will also occur as a result of a fast stop in which the tractor brakes are stronger or find better traction than the trailer brakes.

In rounding a curve the tractor rear wheels will track toward the inside of the curve from the front, and the trailer wheels will track still further in. Swinging turns wide enough to keep the trailer clear and avoiding jackknifing are special steering requirements for forward driving.

Jackknifing is a common occurrence when backing a semi, but is then frustrating rather than destructive.

When a semi is backed around a curve the trailer wheels will move toward the same side as the front wheels, and any difference in direction between the two sections increases as the tractor is backed straight. Jackknifing is more apt to occur if the trailer turns sharply and the tractor slowly, rather than when the reverse is true.

If the swivel point between tractor and trailer is any noticeable distance behind the tractor rear axle, the swivel moves in an arc as the tractor turns and accentuates the steering effect on the trailer.

In backing down a hill the trailer brakes should be released and those of the tractor applied sufficiently to create a drag. The trailer has more tendency to roll straight when pulling than when being pushed.

Full trailers are much more difficult to control than semi's. Most drivers cannot back them more than a few feet.

In general, full trailers should not be used where backing is required. Emergency measures for getting one out of a blind street include unhooking so as to tow it backward.

CHAPTER SIX

GRADERS AND ROLLERS

RUNNING A GRADER. Figure 62 shows a somewhat impressionistic cut of a standard type of heavy duty tandem drive grader.

The grader is normally moved forward while working. It can work in reverse by using the back of the blade for spreading or smoothing; or by reversing the blade so that its cutting edge faces to the rear.

In forward operation the grader is kept in line by steering, and by leaning the front wheels away from the direction of thrust or toward the direction of turn. The wheels may also be leaned to avoid rubbing vertical banks.

Adjustments. The blade is controlled in a number of ways. The ends can be raised or lowered independently of each other, or together. It may be positioned across the line of travel, parallel to it, or at any angle. It can be shifted to the side and into a vertical position by power. There are also mechanical adjustments for extending its range.

The lift arm adjustments are made by removing a pin holding an outer and an inner section together, raising or lowering the lift crank until the set of holes nearest the desired length are in line, reinserting the pin, and locking it.

Manual side shift adjustments are most conveniently made by two men. They are seldom required in open work. The results achieved by the extra reach can often be obtained by better operating technique or by adding a blade extension. Most work can be done with a standard blade centered on the beams, and the lift arms adjusted to within one hole of center.

The blade is ordinarily kept near the

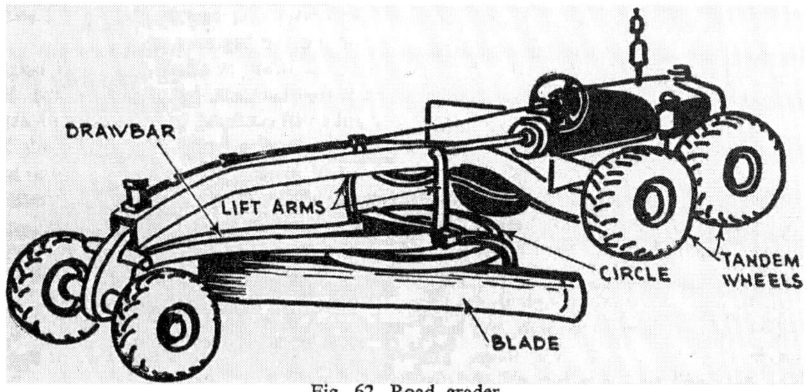

Fig. 62. Road grader

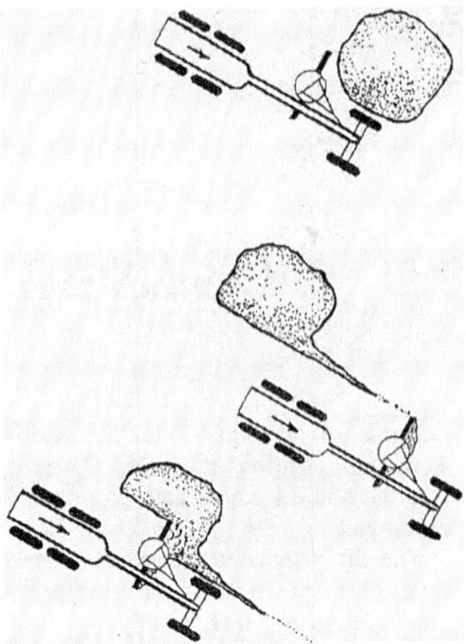

Fig. 63. Spreading a pile

center of the tipping adjustment so that the top of the blade is directly over the edge. Increasing the lean forward decreases cutting ability and causes the blade to ride over its load rather than to push it. It diminishes the likelihood of catching on solid obstructions and may be used for rapid, light planing of rather regular surfaces and for mixing operations. When leaned back the blade cuts readily but tends to let the load ride over its top, and to dig into obstructions. In machines not carrying a scarifier this tilt may be used to cut hard surfaces.

Bulldozing. A grader blade may be used as a bulldozer to a limited extent, often in spreading piles of loose material. If there is space to work beside the pile the blade should be extended well to the side, and the pile reduced in a series of cuts, as shown in Figure 63.

If there is not room enough to do this and the piles are not too high, the front wheels may be driven over them. The front axle will push the top off and the blade will cut as much more as power permits.

The blade should be kept well below its highest position so that if the machine gets hung up on it and loses traction it will be possible to raise it to restore weight to the wheels.

Piles to be distributed by a grader should be spread as much as possible while being dumped.

The grader can also be used for light cut and fill work in building and regrading roads.

The load to be pushed is limited by the power and traction of the machine and will usually be much less than by a crawler of the same weight, although it will be moved faster. The blade itself is quite low, but being more concave than the dozer blade imparts a more pronounced rolling action to the load so that a large quantity can be pushed without spilling over the top.

If the blade is lowered on the right only the other lift arm will remain stationary and the circle and blade will pivot around it, causing the left end of the blade to rise about one quarter of the distance the right side lowers. If the left corner is to be held in position it must be lowered intermittently while the right side is lowered steadily. This effect occurs on either side.

Keeping track of both ends of the blade is necessary and may be the hardest technique for the beginner to learn.

If the blade is raised to its full height and the controls left engaged, the lift cranks will continue to rotate inward until stopped by the frame. The blade will be lowered during the movement from top center to the frame, although the controls are in the raise position. If the controls are moved to DOWN the blade will rise until the cranks are past center after which it will respond normally.

If much dozer work is to be done, it may be advisable to install a front dozer blade.

Side Casting. When the blade is set at an angle, the load pushed ahead of it tends to drift off to one side, as in Figure 64. The rolling action caused by the curve of the moldboard assists this side movement. As the blade is angled more sharply the speed of the side drift increases, so that the dirt is not carried forward as far and a deeper cut can be made.

The sideward movement of the load exerts a thrust against the blade in the opposite direction, which tends to swing the front of the grader toward the leading edge. This thrust is handled by leaning the front wheels to pull against the side draft, and steering enough to compensate for any side slipping which occurs.

The most usual way in which to describe blade setting is to say that a blade set straight across, as in Figure 63, is at zero and all other settings are described by their angular distance from that position. Most road shaping and maintenance is done at a 25° to 30° angle, with straighter settings for distributing windrows and sharper ones for hard cuts and ditching.

The angle of the blade is regulated by the circle reverse control. The mechanism is self-locking and can be turned any desired amount. In some makes it can be adjusted only while the blade is empty or doing light work, in others while pushing a heavy load.

Side shifting the circle from center will raise the blade and change its angle so that compensating adjustments may be necessary.

Planing. If the blade is set at an angle it can be used to plane off irregular surfaces by lowering them sufficiently so that enough material will be cut off the humps to fill the hollows. Enough extra material should be cut to keep a partial load in front of the blade. The forward and sideward movement of the loosened dirt serves to distribute it effectively. If a windrow is left at the trailing edge of the blade it is picked up on the next pass. On the final pass a lighter cut is made and the trailing edge of the blade is lifted enough to allow the surplus material to go under rather than around it, to avoid leaving a ridge.

This type of light planing will produce a smooth surface under favorable conditions, but the fill in the hollows is liable to settle or be compressed below the cut sections. Also the blade may chatter in a very shallow cut, particularly if the mechanism is loose or worn.

A more thorough method is shown in the diagrams in Figure 65. A series of cuts are made across the area to a depth sufficient to reach the bottoms of the holes, or at least to two inches. The large windrow of loosened dirt is then spread back evenly over the area.

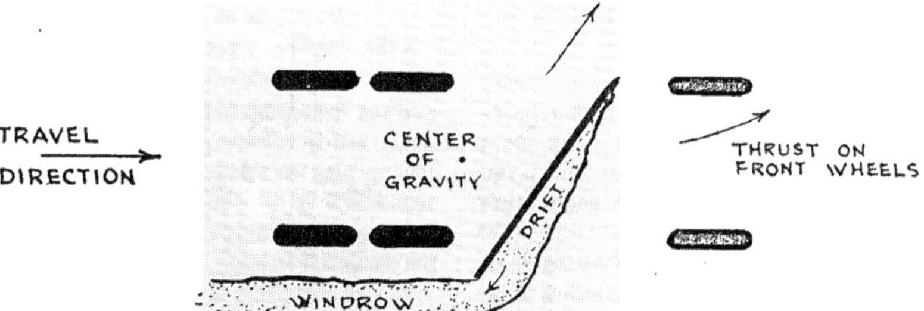

Fig. 64. Side casting

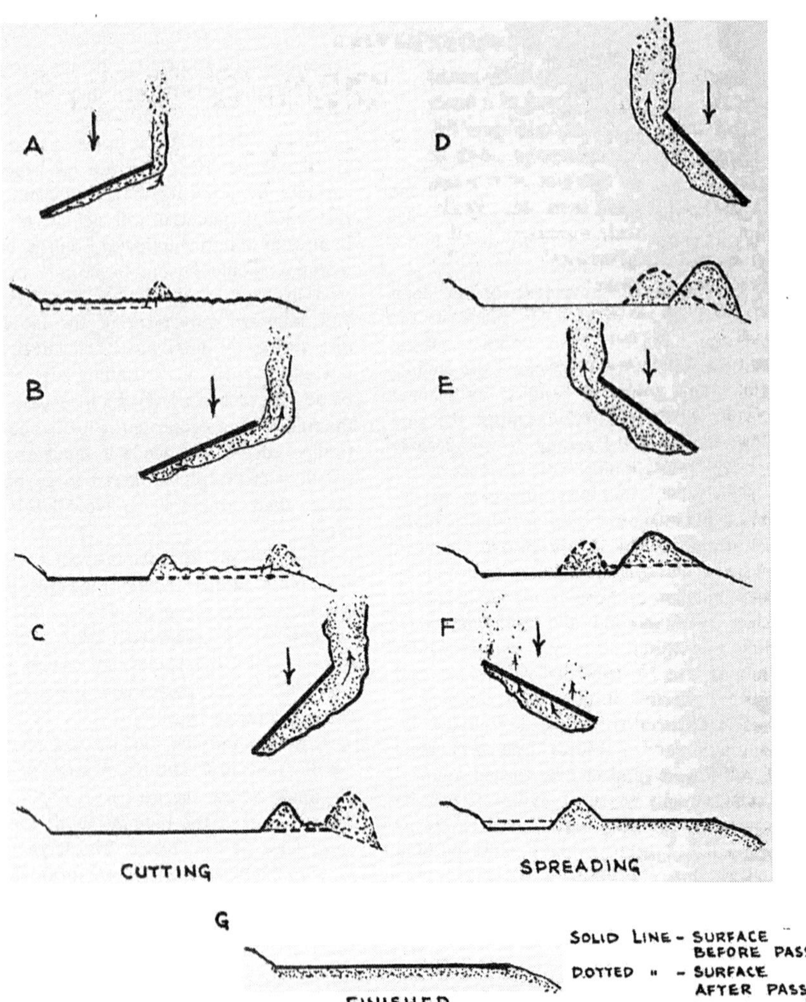

Fig. 65. Refinishing a flat section

It is easier to get a smooth surface with this method because of the advantage of working loosened material and the more uniform distribution with a full blade. The surface will tend to remain smooth after settling or rolling.

It is desirable to vary the blade angle during this work, the first cuts being taken with a straighter blade than the later ones, and the first spreading pass having a sharp angle which will be reduced on each following pass as the size of the windrow is reduced.

Windrows should not be piled in front of the rear wheels as they will interfere with grading accuracy and traction.

Crowning. When the piece to be smoothed is a dirt or gravel road, it is generally crowned so that water will flow off to the sides. Figure 66 shows a sequence of passes in a planing operation. Road material is bladed inward from the shoulders or ditches, and the top of the crown is cut with the blade at zero angle, or at a slight angle which will side cast some material to either side that may require it. The windrows are then spread by putting the blade at an angle of 10° to 25° toward the center, and using a fast working speed. The blade is held above the level of the undisturbed surface so as to avoid collision with solid objects. The speed causes the loose material to be thrown from the blade, so that it will feather out and blend at the top. Any ridge built in the center is then spread out at speed with a straight blade. This should finish the job, but it may be desirable to backblade or rework some sections where proper crowning was not obtained.

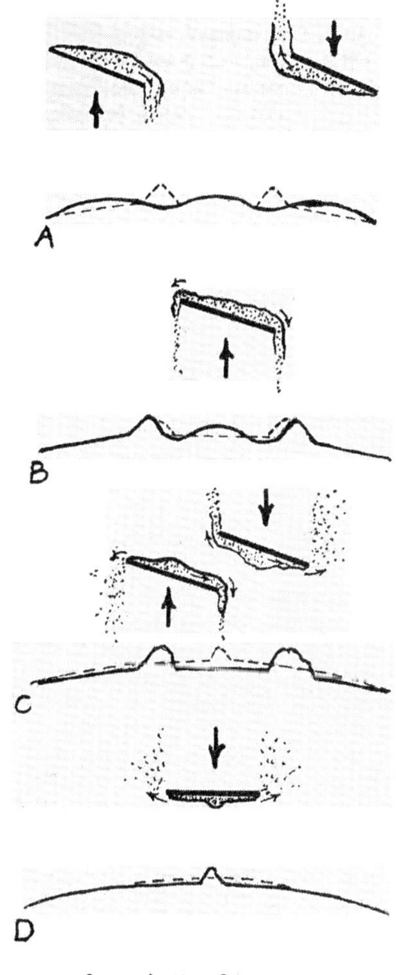

Fig. 66. Crowning a road

If the road is gravel or other imported material and the ditch or shoulder is loam or clay the road may be made muddy by blading in too much from the edges. Since the road must be finished so as to drain off the sides, it may be necessary to blade off high spots on edges outward.

Sods and other debris brought up on the road from the ditches interfere with grading, as lumps catch under the blade, leave ruts, and block the sideward drift of the dirt. If labor or raking equipment is available, it is best to remove the debris from the side windrows before making the center cut. If the grader is working alone, the operator may occasionally climb down to remove a particularly annoying piece, but most of it will be left on the road to dry out and be broken up by traffic.

Stones are a more serious nuisance both in making smooth grades difficult or impossible and in causing damage to machine and operator during cutting. If the cut is shallow or the road dirt compacts readily it is possible to blade loose stones out of the road while grading it.

Large rocks or those firmly embedded in road or shoulders are dangerous. If the blade hooks into one too solid to move, the grader will jump sideward or stop abruptly. This imposes severe shock on the blade, circle, and power train and is liable to throw and injure the operator. Danger of serious consequences increases greatly with higher speeds, so that all cutting should be done in low gear and often at part throttle where buried rocks or roots are expected.

Work Patterns. The following discussion will concern work done by single graders. However, it is often efficient to have two or more of the machines working together, each performing one step in the work sequence. This speeds the job, produces better results than working small sections with individual machines, and reduces or eliminates blocking of roads with windrows.

Road building and grading may be done on three general patterns. One is to work the two sides alternately, turning at the end of the strip. Another is to do one side at a time, working in both directions by means of a reversible blade. The third is doing one side at a time with the reverse trips non-working or utilized for light work with the back of the blade.

The pattern used is determined by the length of the strip being worked, the turning space and footing, and the reversibility of the blade. The machine does its best work going forward, and this increased efficiency must be balanced against the time, labor, and risks involved in turning. In a long run even difficult turning conditions may take an insignificant part of the working time, where easy turns may not be justified on a short run. A general rule is that a grader should not be turned if the strip is less than 1,000 feet long.

Turns. In turning, the front wheels are leaned all the way over in the direction the front of the frame is to turn, and left in this position for both forward and reverse movements until the turn is completed. If ditches or rough ground must be used the machine should be backed into or across them. The oscillating tandem drive will readily climb ditches or obstructions which would be difficult to cross with front wheels.

Turning may be a serious problem. Because of lack of a differential the rear wheels drive straight forward and the tandem arrangement gives them considerable resistance to sideward movement. Weight on the front wheels is light and even when leaning properly the tires may not have enough traction to turn the machine on loose material, but will skid and slide sideways.

If the front tires do grip enough for a sharp turn (minimum radius is about 36 feet), the rear wheels on the inner side of the curve must spin or those on the outside drag enough to compensate for the different distances they travel. Turning a grader sharply therefore results in scuffing and gouging which may be sufficient to damage soft surfaces.

It is therefore often necessary to make a gradual turn either by swinging in a wide circle or by jockeying backward and forward in order to avoid damaging or loosening the turn spot. It is sometimes possible to do most of the turning on ungraded areas where no damage will be done.

Other factors affecting the choice between turning and working in reverse may be idle travel distance between the end of the work and the first possible turning area, interference with traffic while turning, and chances of getting stuck.

Tandem drive affords powerful traction, but a careless operator may hang the blade or circle up on ridges or rocks and the front wheels particularly may sink into mud with the same result. Recently filled ditches which have become water-soaked make grader-traps. Repeated sharp turns on sandy soil may loosen it enough so that the rear wheels will spin in it.

If the lifts are in good condition, many graders can be unstuck by putting blocks under the blade and applying down pressure. This will normally lift the front wheels off the ground so that they can be shored up and the blade then lifted clear. If the pressure is applied to the rear corner of a sharply angled blade, it may lift the rear wheels on one side. The scarifier will lift the front only.

Large front tires—the same size as on the drive wheels—give better front end flotation, are more effective in climbing ditches and banks and in holding the machine to its course on slopes and turns.

Reverse Blading. When the blade is turned so that it will cut at an angle in reverse gear, any heavy load will tend to push the rear wheels sideways as they can neither be steered or leaned to resist the thrust. Reverse work should be arranged so that these wheels can be kept against a bank that will prevent them from sideslipping. Therefore, when doing ditch work in both directions, after the first cut made going forward, the cuts from the bank are made in reverse and the spoil moved away and spread on the forward trips.

Older and smaller models of graders may not be able to turn the blade to cut in reverse. Many of the newer ones cannot be reversed unless teeth are removed from the scarifier, and if the soil were hard enough to require loosening, the reverse position of the blade would not be available until the loosening was finished. While the change can be made quickly in some machines, in others maneuvering the blade into reverse position may take more time than would be justified to work a short trip.

The back of the blade is also effective for smoothing loose material, and can sidecast limited quantities of it. It can be safely used in fast reverse as it will ride over obstructions rather than digging into them. During the cutting and transporting passes back blading is not effective, but it is useful in finishing.

Road Building. A grader, without assistance from other machines or hand work, can shape up a road across a field by digging a pair of parallel ditches and using the soil to build the road crown. However, sod can make the finishing operation tedious and unsatisfactory as it tends to ball up under the blade and catch and pull out of loose surfaces. For this reason, the strip should be thoroughly disked before grading is commenced.

The outer ditch lines are marked by stakes or by the edge of the disked strip. The first cut on each side is made about two feet inside the edge, as in Figure 67 (A). The blade is held at a very sharp angle, perhaps fifty or sixty degrees, with the leading edge just outside the wheel track and the windrow rolled off under the grader. The cut is a light one made primarily to mark the edge of the work and to hold the wheels against side slipping.

The next cut is made at about a 25° angle, as in (B), casting the spoil beyond the inner wheels. If the windrow is large enough it is spread toward the center, as in (C). Otherwise, additional ditch cuts are made until sufficient material is piled to justify a spreading pass.

Ditch cuts, alternated with casting or spreading, are continued until the ditch is the proper depth. The outer slope is then cut, as in (D), and the spoil moved up the inner slope, as in (E), from where it can be spread over the road.

The other side is done in the same manner and the fills are blended at the top. Ditch cuts, except the first one or two, can be made either in forward or reverse. Light casting and spreading can be done in either direction but forward

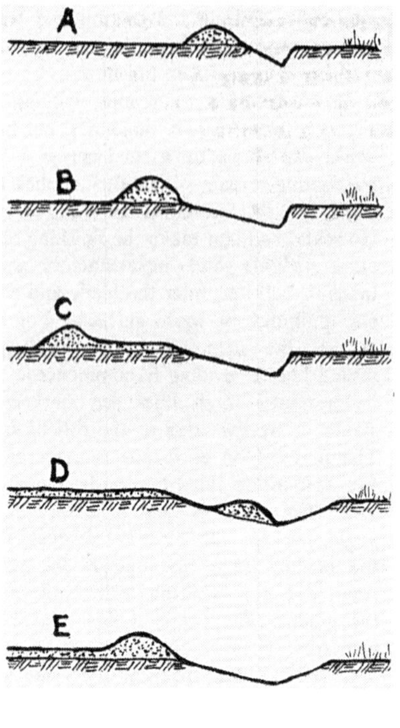

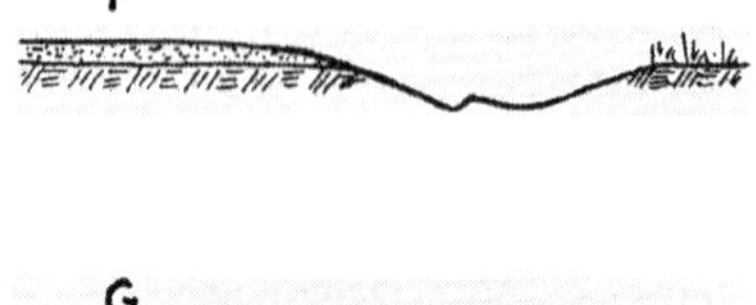

Fig. 67. Ditching for fill

is more efficient if the windrow is heavy.

Manufacturers recommend doing forward ditch cuts and other heavy grader work in second gear at a speed of three or four miles per hour. Blading windrows and similar handling of loose soil can often be done in third gear at speeds up to six miles. However, when there is loose rock, lower speeds will pay off in improved quality of work. In the presence of buried obstructions heavy enough to stall the grader, very slow movement may be required for protection of both the operator and the machine.

If a wide bottom ditch is required the further operations shown in (F) and (G) are undertaken. Slices are cut from the inside slope of the ditch, cutting down to the level of the original bottom and leaving a ridge. This is cut out by running the grader with its outer wheels in the original bottom, setting the blade with its leading edge even with the outer tires and at a sharp angle so that it will cut only the width of the desired flat bottom. The spoil is piled on the bottom of the inner slope from where it is cast onto the road and spread by following passes.

The number and sequence of passes are affected by the depth of the ditch, the width of the road, and the resistance of the soil.

STEEL WHEEL ROLLING

Controls. So far as the controls are concerned roller operation is simple. In certain machines the reversing clutch lever is put in neutral, the shifting lever moved into the desired gear, and the clutch lever moved forward to roll forward, back to move backward. In some other rollers the main clutch is used for shifting and the reversing clutch for choosing direction and for starting and stopping.

Hydraulic steered machines are turned by moving the steering bar in the direction of turn until the front roll is at the angle desired. The bar is returned to center until the turn is completed, moved to the other side until the front roll is straightened out, and is then returned to center. Any adjustments in sharpness of turn may be made by moving the bar one way or the other.

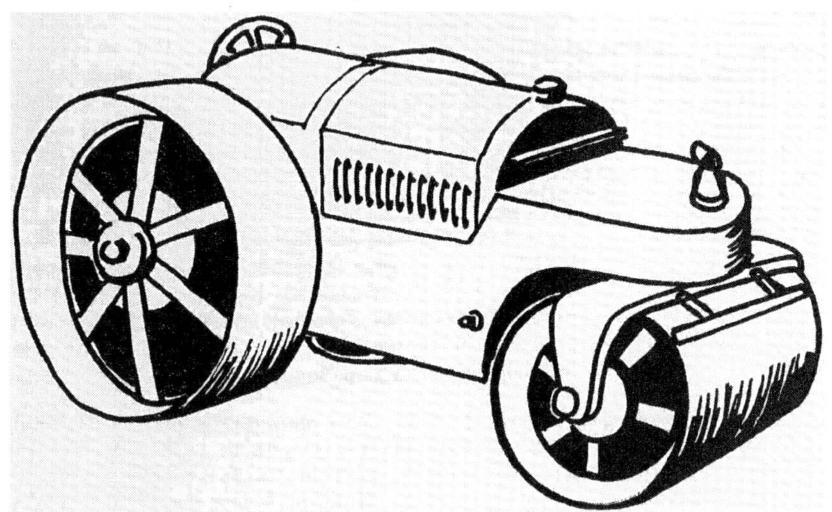

Fig. 68A. Three wheel roller

The error to be avoided is leaving the bar in the side position as one would a car's steering wheel. This causes the ram to continue to move, making the angle sharper and sharper until the stops are reached.

Steering sharply is liable to cause scuffing and may result in severe damage to the surface when the roller is moving backward. A sharply turned front roll has a tendency to dig in on the edge turned toward the rear, and then tip so as to dig in further so that a rut is plowed in the surface. Work sequences should be arranged so that turns can be made while moving forward.

The differential lock is not used during compacting work. It is intended to give additional traction when scarifying, or walking over rough or steep ground.

The sprinkling system cannot be used in compaction of subgrades as dirt or gravel will stick to the wet rolls and come up in chunks, spoiling the surface. If water

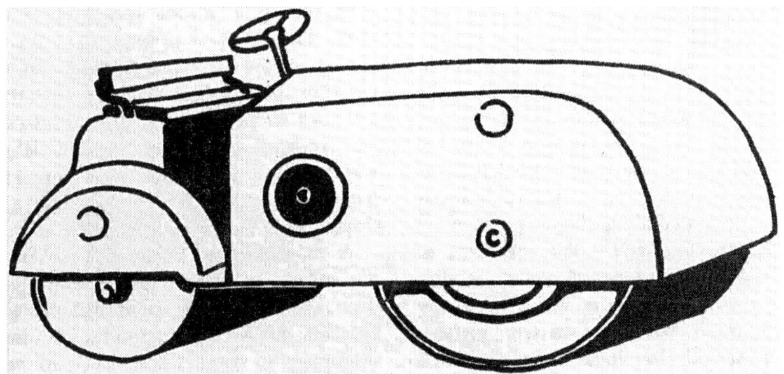

Fig. 68B. Tandem roller

Roller Faces ←

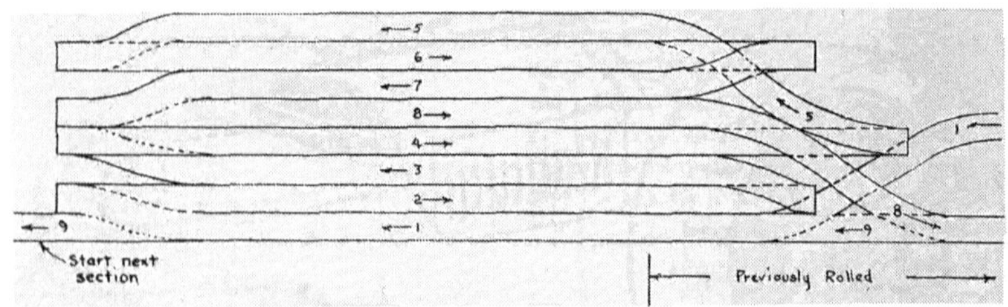

ROLLING SEQUENCE ON CROWNED ROAD
overlapping not shown

Fig. 69. Rolling sequence, straight

is needed for the compaction it should be supplied by water wagons sufficiently in advance so that it can dry slightly before the roller reaches it.

Stopping. Stopping and starting should be done gradually in order to avoid scuffing of the surface. It is best to disengage the clutch before reaching the end of the run so that the roller can drift to a stop without using the brake or reversing the clutch. When it is stopped, the clutch for movement in the opposite direction is gradually engaged for a smooth start.

The roller should not be stopped repeatedly in the same place, particularly on blacktop, as this may cause creeping and formation of pockets.

Obstructions. Manholes or other obstructions in the road interfere with regular rolling patterns. The way they are handled will vary greatly with their height, construction, and location. In rolling up to an obstruction head-on, the clutch should be released or partially released before coming to it, so that the roll will just touch it without having momentum enough to break or climb it.

The pattern is rolled the first time as nearly normally as possible and curving passes made later. Spaces which the roller cannot reach without excessive maneuvering should be hand tamped or a smaller roller should be used in them.

If the obstruction is both low and strong it may be possible to ride the front roll over it and straddle it with the rear rolls.

Sequence. Rolling speeds are slow. One and a half to three miles an hour is usual.

The rear wheels of a roller do most of the compacting, particularly on the three wheel type. The smaller and lighter front roll serves to "work" and stabilize the soil.

In rolling deep loose material such as fill or gravel all passes in a series except the first should be overlapped at least half the width of the drive roll. Gradual extension of the rolled into the unrolled area makes possible greater concentration of weight on local ridges and high spots, and keeps the rolls running at a truer grade.

In rolling a graded area with a side slope, as a crowned or banked road, work should always be done from the bottom up. The lower edges of the rolls have a tendency to push downhill, which can be best resisted by compacted material. In working uphill the creep of soil away from the upper edge helps to preserve the slope.

A crowned road is rolled according to the pattern in Figure 69, starting at one edge and working up until the center is reached by the upper roll, then moving diagonally to the opposite side and working up from there. Each re-rolling is started at the bottom in the same manner.

It is efficient to roll sections as long as are ready for work at one time. Two procedures in general use are shown in Figure 70. In (A) a section is done completely at a time with ends overlapped a few feet, and in (B) a short section is started and lengthened steadily in one direction, following grading operations. When the original piece appears to have had sufficient rolling it is abandoned and the strip being rolled kept at one length afterward, being extended at one end and shrinking at the other.

Banked or super-elevated curves are rolled from the inner edge to the outer edge, as in Figure 71. The transition from crown to bank is made by a diagonal from center to low side. From bank to crown the move is from either edge straight into adjoining low side. The meeting of these two types of grade is a convenient place to end a rolling section if the continuous advance system is not being used.

Rolling should be continued until no advantage is noted from successive passes. Presence of too much water in the subgrade may make its compaction impossible, but long rolling will at least bring much of the water to the top where it can evaporate more readily. The waterlogged condition results in a rubbery action of the ground, in which it goes down under the rolls and springs back into nearly its original condition when they have passed. This condition may not be apparent at the start of the work, as the larger air spaces in the unconsolidated soil may be adequate to hold the water. As these spaces are reduced, however, the water is forced out of them and becomes a lubricant between all the particles.

Scarifying. The scarifier is used principally in ripping up old blacktop or oiled surfaces, but can also be used to pull out

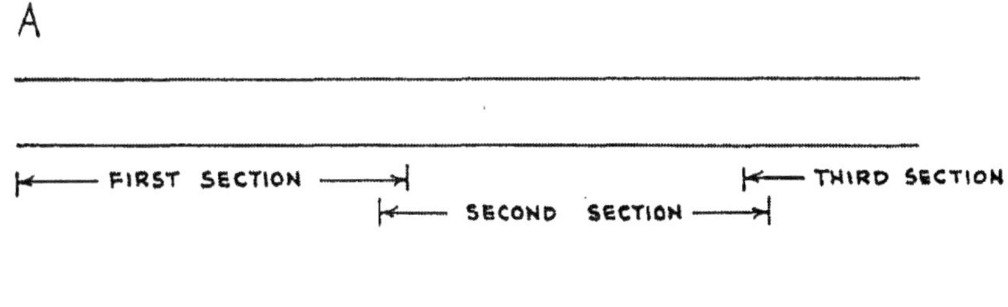

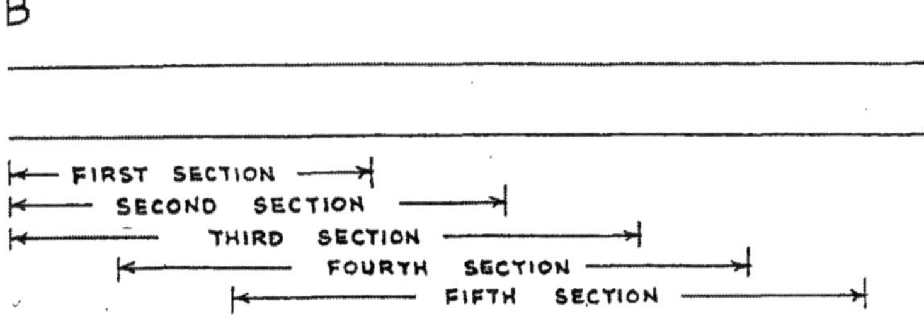

Fig. 70. Overlap of rolling sections

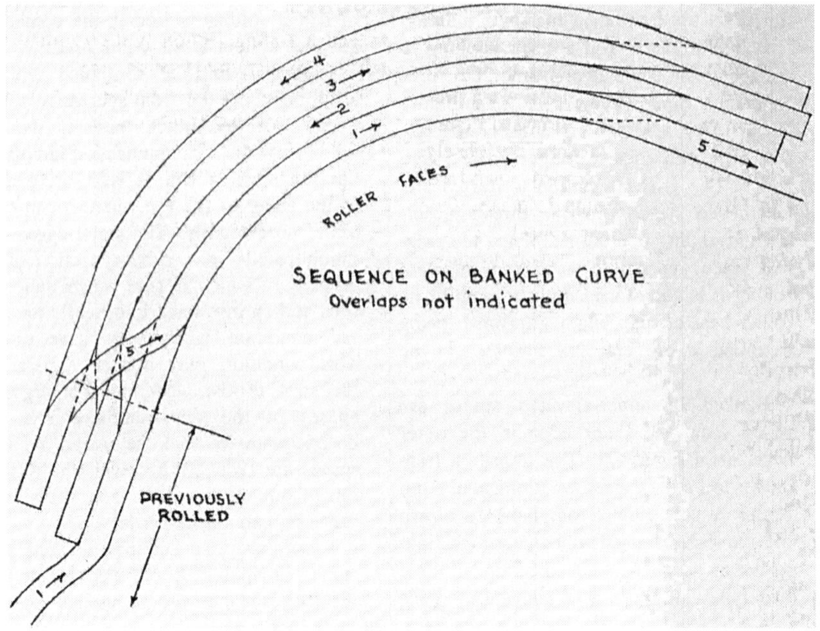

Fig. 71. Rolling banked curve

rocks or to loosen hard soils for a grader.

A roller does not have much traction on hard surfaces, so under many conditions it does not have enough pull to scarify effectively. This may be partially compensated by working downhill, by shallow cutting, and by reducing the number of teeth.

If a heavy pull is required, the drive rolls may be fitted with spikes. Tapered plugs are driven out of the rolls and points substituted. This is a laborious operation unless the rolls are new or spikes have been installed frequently. With worn rolls, there may also be some difficulty matching the plugs into the holes when replacing them, so that the smooth surface will not be spoiled by pits or bumps.

The actual scarifying is done by locking the roller differential, lowering the teeth into the ground or pavement by hydraulic control, and driving forward in low gear. At the end of the pass the roller is turned if the strip is long, or backed to the beginning if it is short.

If a pavement breaks into pieces too large to pass between the teeth they will become plugged and the scarified material will pile up ahead of them, as in front of a dozer blade. When this occurs the scarifier should be lifted and forward motion continued until the teeth have passed over the pile. They are then positioned so that they just clear the pavement. The roller is backed and the pile pushed backward by the teeth until it is cleared off the unbroken pavement.

The teeth are then lowered into the pavement at the nearest point where full depth penetration was obtained, and the scarifying resumed. This cleaning off may be done as often as it is necessary.

CHAPTER SEVEN

OTHER EQUIPMENT

AIR DRILLS. Hand Drilling. The rock drill is carried to the work area and connected to a flexible air line from the compressor. A short starter steel, a foot to thirty inches in length, is fitted with a bit and clamped in the chuck. This bit should be the largest size which is to be used in the hole.

The area surrounding the hole should be free of dirt and litter that would interfere with starting the hole or might slide into it during drilling.

If the rock surface is horizontal, the drill is held vertically and the throttle partly opened. If the bit wanders on the surface instead of cutting it, it can be held in position with a foot pressed against the lower end of the steel. If the rock slopes the steel should be held at a right angle to the slope instead of straight up and down until it has dug a pit that will hold it.

When the bit has cut in enough so that it is not likely to bounce out the hole is said to be collared. The throttle can then be fully opened. The drill can be simply supported against leaning over, or if it tends to bounce, it can be pushed down with one or both hands, or (over manufacturers' protests) by hanging a leg over it.

In soft rock it is not advisable to push down on the drill as it increases any tendency toward rifling or binding. It is sometimes necessary to pull up slightly to avoid these difficulties. In hard rock extra weight is helpful, particularly if the drill is light.

While the hole is shallow the automatic puff blowing should keep it clean. However, it is good practice to turn the valve to blow occasionally. The amount of extra chips and dust raised will indicate whether the puff blowing has been doing the job.

When the hole is deepened so that the drill is close to the ground, the throttle should be closed and the drill lifted until the bit and steel are out of the hole. The steel is taken out and another steel about two feet longer is locked in the chuck. If it is not much worn, the starting bit may be put on the second steel.

The bit is detached by holding the steel firmly and striking the bit upon the wings so as to unscrew it. The steel may be held under the foot. Some operators prefer to use wrenches—a pair of Stillsons, or special wrenches—for changing bits. In installing a bit, it is hand tightened then tapped to seat it firmly.

It should be remembered that the thread is left hand.

If another bit is used, it should be slightly smaller than the starter. The easiest way to compare sizes is to put the two bits together.

Drilling is resumed with the second steel. When it is mostly in the hole, it is replaced with a longer steel, using the same or a smaller bit. This process is repeated until the hole is finished. Its depth is readily measured from the length of the steels.

As the hole deepens, it becomes increasingly necessary to blow it out. Some hammers can be adjusted to send down an extra stream of air while drilling, which reduces the need to stop drilling to blow.

If air and dust stop coming out of the hole it should be blown immediately. If it will not blow the steel should be pulled out. The trouble is usually a plugged bit. The cuttings can be removed with a thin punch, a nail, or an ice pick.

An attempt to drill without a flow of air will overheat and spoil the bit and probably cause a collar of compacted cuttings to form just over it, which will make it very difficult to pull.

Whenever possible, holes should be drilled vertically. This puts the weight of the hammer on the bit instead of on the operator and simplifies keeping the hole straight.

Jamming. There are several conditions under which getting the steel and bit out of the hole are a greater problem than the drilling.

Generally, the principal cause of sticking is the use of worn bits. In most ground the bit gives a warning that it is reaching a danger point before it jams. This may be slower penetration, production of fine rock dust instead of chips, twisting of the drill in the hands, or failure of the steel to rotate. Condition of the hole may be roughly checked by raising the drill a few inches occasionally to see if it is free.

In hard or brittle rock a bit may function satisfactorily which is so worn that it would stick immediately in a soft formation. However, worn bits are also dull bits and drilling is slowed by their use. It is also poor economy to use a worn bit because excessive wear may make it impractical to recondition it, particularly by grinding.

In soft rock a bit may rifle the hole. This trouble may be expected whenever the drilling rate is very rapid. New, sharp bits are worse offenders than worn ones. Precautions which can be taken include use of special X or rose bits, using a lighter drill or partly supporting the weight of a heavy one, grinding back the cutting edges of the bit, or reconditioning bits by grinding the sides to restore taper without sharpening the edges.

Seams of dirt or finely disintegrated rock in a formation will cause plugging of the air hole. At the same time there is rapid penetration and dirt falls on the top of the bit, where it may compact into a hard, tight collar.

When the speed of penetration increases suddenly drilling should be stopped immediately and the hole blown. Drilling can then be resumed on part throttle with frequent, thorough blowing. If air stops coming out of the hole the steel should be pulled and the hole in the bit opened.

If the ground is composed of sloping layers of rock of varying hardness, or seams or cracks slope across the line of drilling, the hole will tend to drift down the slope. The curved hole that results will bind the steel.

Drilling on a dip at right angles to the seams should prevent this trouble. Often an intermediate direction will be satisfactory. In vertical drilling, curving can often be prevented by using light pressure and sharp bits and arranging that the drill gets maximum air pressure. This may involve cleaning the drill air filter, shortening and straightening the line from the compressor, or temporarily shutting down other units using air from the same source.

Pulling Steels. In taking a steel over three feet long out of a hole, the usual procedure is to unclamp the steel puller, set the hammer aside, and lift the steel by hand. However, if the steel is stuck, the hammer should be left attached and lifted with the throttle alternately in drilling and blowing position. The heavy vibration and rotation of the steel and the air pressure help to break it free.

Two men—one on each handle—can lift the drill much more effectively than one.

If the sticking is caused by spiraling (rifling) it may be possible to extract the steel by turning it against the direction of rotation while lifting. If two men are working, one can lift the drill with throttle closed while the other turns the steel and drill with a wrench. If the operator is alone he should remove the drill, place the wrench just under the shank collar, and pull the wrench upward while turning it.

This may cause the steel to unscrew from the bit. If the hole is full depth a steel bit can be sacrificed economically. If it is not full depth, leaving the bit in will make it necessary to drill another hole. If the bit is carbide, every effort should be made to salvage it by direct pull.

Usually steel and bit can be raised together by jacking. No really suitable jack is generally available, but a chain jack or a regular bumper jack with a chain grab fastened to the lifthook can be used. The jack is placed on the edge of the hole, the end of a light chain given a double turn around the steel and hooked, and the chain pulled tight and caught in the jack head. The best grip is just under the collar, but a small wood wedge will generally enable it to hold anywhere on the steel.

This jack does not pull straight up and jack and steel will tip as the steel rises. This may cause bending of the steel or the jack if lifting is continued too long on one hold. It is a good plan to release the chain and fasten it again further down before bending is severe.

A chain fastened to wagon drill feeds, loader buckets, bulldozer blades, cable control units, or to any type of shovel or crane may be used for steel pulling.

If all methods of extraction fail a parallel hole is drilled nearby and the steel picked out of the loosened rock after the blast. It is generally bent by the explosion, and unless it lands on top, may be further twisted by a shovel.

Care of the Drill. A drill is built of special steels and is very finely machined throughout. It costs about five times as much per pound as a shovel or dozer, and it deserves much better care than it usually gets.

The oil reservoir is built into the bottom of the cylinder and may require filling from one to four times a day, depending on the make and condition of the drill. This oil is forced or sucked into lubricating passages, and into the air stream during work. The oil mist in the exhaust air gives an indication of the amount being used.

A line oiler should be used instead of or in addition to the reservoir for best results, as it has a larger capacity and is more reliable.

The frequency with which either a reservoir or an oiler must be checked is best determined by experience. If exhaust air ceases to carry oil, the cause should be found immediately.

Air drills have very slight clearances and have a severe temperature drop. Air often enters them at a temperature of a hundred degrees or more, and is exhausted close to freezing.

In addition, water condensing from the air or leaking from a water tube may wash oil off the parts which need it. Special oils designed for these conditions will give better service than standard lubricants. However, any oil is better than no oil.

Every possible precaution should be taken to keep the oil clean. This is a difficult matter when the air is full of rock dust, but it is important. Dirt may clog oil or air passages or score sliding parts.

Use of an inlet air filter may save trouble with scoring or clogging, but it should be cleaned frequently. Most filters are so small that a little debris can reduce the air passage enough to starve the drill.

A drill needs periodic cleaning, the interval depending on the quality of air it receives. An oil-pumping compressor, very hot air, or deteriorated hose may foul it in a day or less. With clean air it may function well for weeks.

A quick cleanout can be done by disconnecting the air hose, pouring a cup of kerosene or fuel oil into the drill, reconnecting the air, and running the drill idle for about a minute. The hose is again disconnected, a cup of rock drill oil poured in, and the drill operated again.

If it is very dirty, several doses of a mixture of three parts kerosene and one part oil can be used.

At much longer intervals, it is good practice to take the drill to the shop (drill doctor), to be disassembled, thoroughly cleaned, and checked for worn parts.

WAGON DRILL OPERATION. A standard wagon drill, Figure 72, consists of a wheel mounted main frame or chassis, a support which can be raised and lowered, and a mast or guide shell. The drill is mounted on a slide on the mast and is moved along it by a roller chain driven by an air motor.

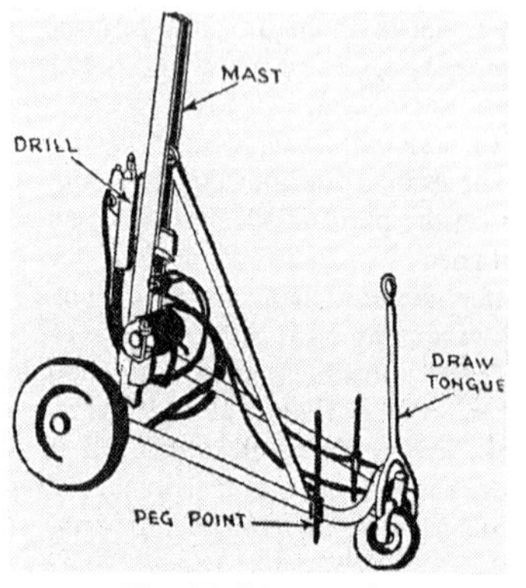

Fig. 72. Wagon drill

The major problem in wagon drill operation may be getting it in position to work. Its weight of six hundred to two thousand pounds, and its small wheels, make it difficult to maneuver.

On level ground it may be shifted by several men. It can be moved over bumps or up grades by putting a steel in the drill, setting the mast at a low angle, and using the feed down pressure to push it, while the draw tongue is held in steering position.

In more difficult places, it may be towed by a tractor, truck, or winch line, or lifted and placed by a shovel or crane.

Planks or light channel irons can be laid on the ground to serve as tracks for the wheels. These can also serve as guides for lining up holes.

When the machine is placed, the mast is put in correct drilling position, the support is unclamped then raised or lowered by hand cranking, hand jacking, hydraulically, or by attachment to the drill feed. If the holes are to be straight down, the support must be high enough to allow the mast to be vertical without striking the ground. Raising it further brings the mast toward the rear, and reduces the effective length of the feed. If the hole is to be close to a face, the main wheels are turned sideward.

If the drill is brought to the bottom of the mast, the mast will hang vertically when its clamps are released. Some masts are pivoted so that they can only swing forward and backward, while others can tip to the side. If the mast will not tip sideward, and one front wheel is higher than the other, the low wheel must be blocked up. It can be raised by putting a steel in the drill, with the drill idle, and pushing it down with the feed. One man can sit on the high wheel so that only the lower one will be raised for blocking.

When the machine is adjusted so that the drill bit contacts the rock at the right spot, the clamps on the U-bar and mast pivots should be tightened. The peg point clamps should be loosened and the pegs adjusted for holding against side slipping in any direction. They are more effective if the machine is raised slightly with a bar before setting.

The two arms of the steel centralizer should be then brought together and clamped on the steel. If the main air line has been disconnected for moving, it is re-attached, and the air valve opened.

The feed control is turned to down position. As the bit touches the ground, the drill throttle is opened. Until the hole is well collared, feed pressure should be light. The centralizer arms are then unclamped and swung out of the way.

The handling of the feed control depends largely on the characteristics of the machine and the rock. The object is to apply enough pressure to keep the bit in good contact with the rock, without rifling the hole or lugging down the drill.

If there are separate speed and pressure controls for the feed, the speed can be set for a faster rate than actual penetration in hard rock, and the pressure control used to prevent crowding. In soft rock, the feed should move slowly enough to avoid rifling.

The drill is operated on full throttle. Puff blowing should keep the first few feet of hole clear. When it proves inadequate, a combined drill and blow setting can be used, supplemented by occasional full blowing.

When the drill reaches the bottom of the mast, it is put on full blow until no more dust can be raised, then shut off. The feed is reversed and the drill raised. The steel is detached and put aside. The next longer steel, fitted with the same bit or one slightly smaller, is put in the hole and the drill raised or lowered until the shank can be clamped in its chuck. The bit is lowered to the bottom of the hole with the feed, and drilling is resumed.

The second steel will be too long to be pulled out of the ground by raising the drill. It is raised far enough to free it from the bottom, then released from the drill and pulled out by hand. A loop on the top of the mast can be used for steadying a long steel as it is raised and for stacking steels not in use.

If the steels are too long and heavy for hand raising, the drill is run almost to the top of the mast with the steel attached. The steel is wedged against slipping back, the drill released and brought down as far as possible without bending the steel. The drill is then tied to the steel, raised, and the steel wedged and untied. This process is repeated until the steel is free and can be swung away from the hole.

If the hole is to be drilled at any angle other than straight down, the wagon is placed facing the direction of drilling. The peg points should be set to resist backward thrust.

Under ordinary conditions, the bit should make hole between 8″ and 36″ a minute while drilling. Time lost in changing steel and bits may cut progress more than 50 percent.

RIPPING GROUND. A heavy rooter may be used with all three teeth, with the center tooth only, or the side teeth only. Use of all the teeth produces most thorough breakage of the soil and works a strip approximately as wide as the towing tractor. Use of the outer teeth only produces coarser pieces, requires less power to pull, or can be pulled at a greater depth with the same power. A center strip may be left unbroken.

A single tooth breaks a narrow strip coarsely and takes minimum power. It is

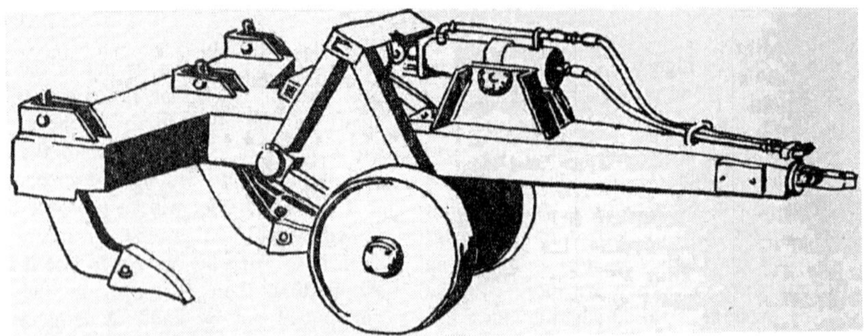

Fig. 73. Rooter

ordinarily used for very resistant material. If full breakage to depth is required, it may be necessary to run successive cuts so close together that the tractor will walk on the previous cut each time. If the soil is wet or plastic, this may compact it so that much of the advantage of the ripping is lost.

Dislodged boulders may make it impractical to keep close along the previous cut unless the tractor has a usable dozer blade.

When breaking heavy pavements, frozen dirt, or other strong crusts overlying comparatively soft material, it may be necessary to run the teeth under the edge, stop the tractor by disengaging the steering clutches, and lift until the pavement is broken. The teeth can then be lowered and pulled forward until solid pavement is encountered and lifted again.

Large slabs, boulders, or stumps may overturn the rooter or jam between the teeth and the wheels. Manipulation of the hoist, together with moving and turning the tractor, will often serve to right and disentangle the machine. Another tractor or heavy truck can right it with a side pull, but if none are at hand the tractor may have to be disconnected.

Overturning is most likely when working with a single center tooth, and least probable with the two side teeth. Jamming occurs most often when three teeth are used.

Except when breaking hard surfaces (pavement and frost) over soft material, teeth should be near full depth. Shallow ripping is irregular and wears points excessively.

Turning while the teeth are in the ground is apt to break the shanks.

Mechanical Lift. Towed rippers can also be equipped with a mechanical or ground wheel lift, similar to that used on agricultural plows. Such machines are called panbreakers, subsoil plows, or subsoilers.

The lift mechanism usually consists of a hook or dog which can be caught on a rack or gear on the wheel so that rotation of the wheel raises the frame to transporting position. Releasing a catch will allow it to drop. Working depth is regulated by a threaded crank. Control is by trip rope. Some models have a rope for raising and another for lowering, while others use successive pulls on the same rope for the two operations.

Some of these machines are built to carry only a single shank or standard. Others may carry one to three heavy standards, or up to five lighter ones (chisels). Maximum depth of penetration varies between 20 and 36 inches, and horsepower requirement from 20 to 90.

They are designed primarily for agricultural work. They are also useful to the excavating contractor, who can use them for breaking hard soils on jobs too small to justify bringing in a heavy duty rooter, for shallow ditching, and for subdrainage. Their cost is comparatively low, and since they do not rely on cable or hydraulic controls, they can be towed by any machine with sufficient pull. However, they are not efficient in pavements or frost heavy enough to require upward breakage, as the lift does not operate except when the wheels are turning, and too much vertical resistance will cause them to skid.

LOADING A DECK TRAILER. It is standard practice to walk machines onto deck trailers over the back with the aid of ramps, blocks, or banks, or combinations of them.

Ramps. Ramps may be made of planks of oak or other strong wood, two to five inches thick (three is usual), and ten to sixteen inches wide. A metal angle is fastened under one end which rests on a shelf on the trailer.

Ramps may also be made of steel box or channel construction, or of wood reinforced with steel plates or angles.

They are ordinarily so heavy that it takes at least two men to handle them, so that it is often advantageous to place them with the machine that is to be loaded. However, they are usually not strong enough to take the weight of a heavy machine unless supported between the ends. Short pieces of heavy timbers, railroad ties, or heavy saw horses can be used to block them up.

In loading crawler machines, the point of greatest strain is near the ground where the tracks are forced upward into a climb. In unloading, bracing is particularly needed where the machine strikes the ramp as it pitches off the trailer deck. In addition to these special points, it is good policy to have blocking under one or two more points to reduce the length of unsupported spans.

Thorough blocking will make possible the use of lighter ramps and add materially to their life.

Ramps supplied with a trailer are designed to load most machines on level ground. Length should be at least three feet for each foot of trailer height. The incline should be considerably less than the machine is capable of climbing.

Rubber-tired machines impose a concentrated but fairly even strain on ramps. Rollers must have very slight gradients and need help from dirt piles or banks to get on a deck trailer.

Crawlers may be loaded either forward or backward, but are under better control when backing uphill, particularly if the tracks are loose.

Ramps are set to line with the tracks or wheels of the machine, which is placed so that it can move up them without turning. If the ramps are wider than the tracks, they are set so that the operator can see an edge and steer accordingly. If they are narrow so that he cannot watch them, a competent man should stand on the trailer or the ground and guide the operator by signals.

The machine should be in its lowest gear and should move slowly, but with sufficient throttle to avoid any chance of stalling. It should not be steered sharply nor stopped except in emergency. A loading sequence is shown in Figure 74.

When the top is reached, the upper end of the tracks will move upward above the tires and deck, and fall when the machine overbalances. A jarring fall can be avoided by moving very slowly. A sloping extension on the rear will provide an intermediate grade between the ramp and deck which reduces the abruptness of the fall.

If the ramp is short it can be extended by blocking, as in (E).

Wet ramps are dangerous and should be sprinkled with sharp sand. Steel sur-

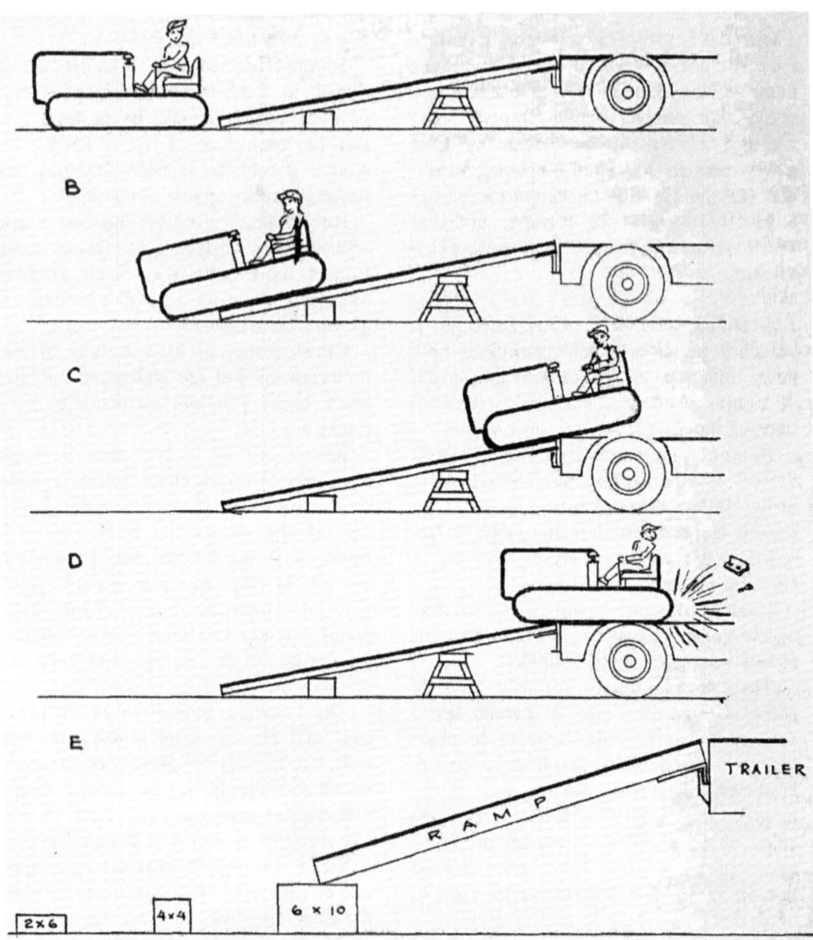

Fig. 74. Climbing a ramp

faces may be too slick to climb even when sanded. Machines should never be loaded up ramps when there is snow or ice on the ramps, the deck, the tracks, or the ground where the machine is standing.

Since the machine must walk over the tires after leaving the ramps, tracks should be inspected and any sharp pieces of metal removed. Grousers will not damage tires, but some varieties of ice cleat might when new.

If the trailer can be backed against a bank, or into a deep gutter, it may be possible to load it without ramps, and frequently even without blocks. Figure 75 shows four situations in which ramps are not needed. Lower banks or shallower gutters may be useful in making it possible to ramp up at an easier gradient than from a level.

These trailers are also used extensively for carrying bulky objects such as tanks, machine parts, small buildings, and pipe.

Picking up and removing "dead" loads, including disabled machinery, is greatly simplified by a rear mounted winch on the tractor.

Pieces which are wider than the trailer may be jacked and blocked up, the trailer backed under them, and the blocks knocked out.

Protection of Pavements. When crawler machines are loaded or walked on paved streets, the road surface must be protected. Damage can be caused by side scuffing during turns, and by digging in of grousers.

Crushing is particularly likely on oiled gravel when the subgrade is wet, or on any blacktop in hot weather. It is guarded

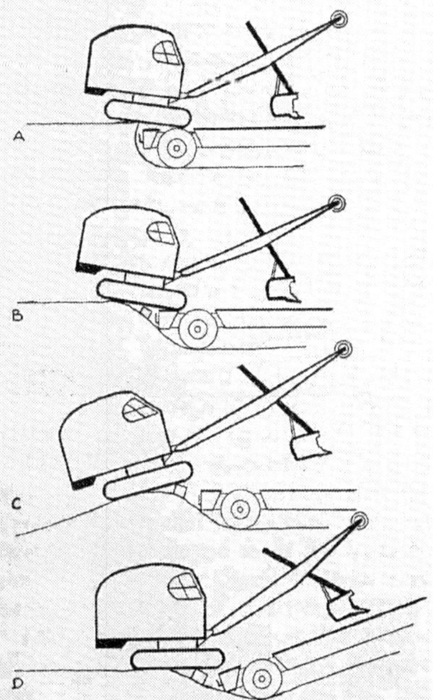

Fig. 75. Loading from a bank

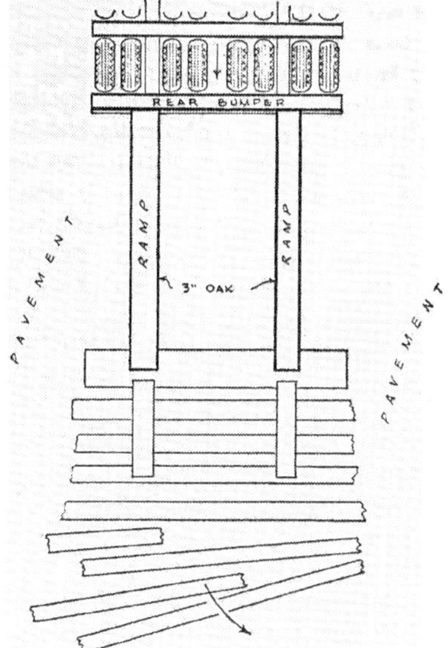

Fig. 76. Protecting pavement

against by laying planks or boards as in Figure 76, to spread the weight over a wider area. Note the extra boards where the front of the tracks tends to dig in just below the ramps.

In many localities, it is unlawful to walk crawler machines on streets but nothing is said if no visible signs of the move are left. Bruising of pavements, and scuffing on turns, can be prevented by laying thin boards under the tracks. A single thin board, narrower than the track, may protect hard pavements, while softer ones may need wide planks, or narrow planks in pairs.

If the walk is long, the same boards can be used repeatedly.

Digging in of grousers may be prevented by walking on boards in the same manner.

Side Loading. It is often more convenient to load from the side. Normal curb and sidewalk heights above the gutter permit use of shorter and lighter ramps. Drop deck models are lower at the side than at the back. In addition, the machine may have to enter or leave the job through a driveway or other narrow access. By side loading, it can be walked directly into the drive, where rear ramps will put it on the street pavement and necessitate a right-angle turn, often under tight conditions. Also, it is frequently possible to pull up alongside a roadside bank where backing into it would not be practical.

Trailers are not ordinarily equipped with ramp-supporting rails on the sides, but they can be welded on easily. There is usually enough recess in the girders to do this without increasing width.

Ramps can be laid on the edge of the deck, as in Figure 77 (A), but this is not recommended since they may be kicked off by the movements of the machine. Ramps can be supported on blocks, (B), or blocks used alone, (C). Short ramps, properly fastened to the trailer, are superior to any of these methods in safety and convenience.

The trailer may tip sideward as the machine climbs on the edge. This simplifies loading from blocks or a low bank. If ramps are used, the extent of tip should be controlled by blocks under the edge to avoid injury to ramp brackets.

The machine can be carried crosswise or turned on the deck. If the machine is large in proportion to the deck space, turning may be difficult or impossible. Machines narrower than the deck can turn rather readily unless the planking is wet and slippery.

The front corners of the deck should be blocked up before turning, but the

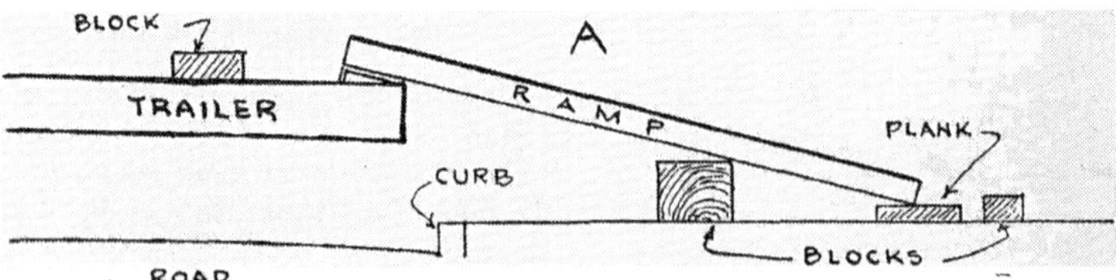

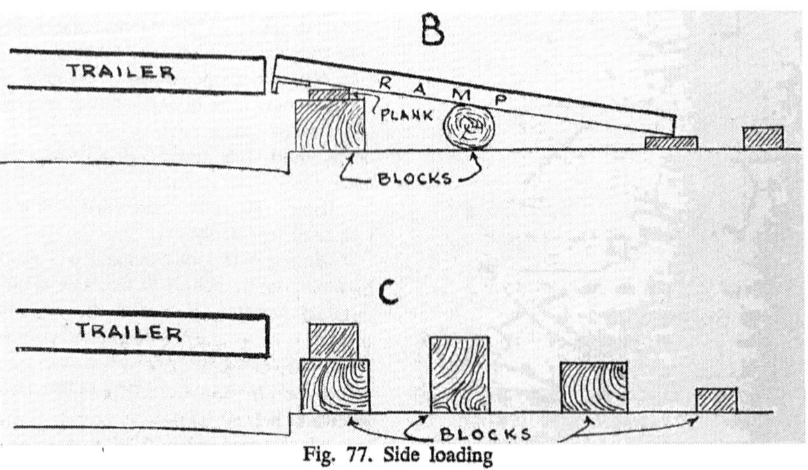

Fig. 77. Side loading

blocks must not be so high that the deck will rest on them when loaded.

Transporting. The amount of load on rear trailer tires, and on those of the dolly or tractor, can be adjusted by placing the machine toward the front or rear of the deck. Placement will also be affected by the need to rest the bucket or blade, and the direction in which the boom is to be carried.

No part of the load should rest over the tires unless high enough to allow them to oscillate without hitting.

A dipper stick bucket may be tucked under the boom on the deck, rested on the gooseneck or between the rear tires, or, if the tractor is a dump truck, in its body. Long booms are usually faced to the rear and carried low, with the bucket or hook resting on the deck. It may be necessary to have the operator stay with a large shovel in transit to raise, lower, or swing the boom to avoid obstacles.

When the machine is correctly positioned, it is left in gear with the clutch engaged and the brakes locked. The ramps and blocks are loaded onto the deck and the machine secured against sliding by blocks or chains, or both.

Chains may be fastened to the four corners of the machine, and to the "D" loops on the trailer sides. At least two of these chains should be tightened and locked with load binders. Sometimes a machine can be lashed by passing a chain from a loop through or over it to a loop on the other side, and tightening it.

Blocks should be placed at each end of both tracks. The ramp blocks or the ramps themselves can be used, but triangular blocks or ones curved to fit the track or tire are much more effective. These may be wedged in place, or nailed down lightly.

Many machines are moved without any chaining or blocking, particularly on short trips. Crawler tractors tend to stay put very well. Shovels, because of their height, are less stable. Rubber-tired vehicles tend to bounce and creep.

Blocking is recommended for all machine transportation, since a minor accident may become a serious one if it slides a heavy unit off a trailer. In some places, truckers are required to block and chain trailer loads.

In most states, truck and trailer widths are limited to eight feet. Wider vehicles must obtain special permits, which may be limited as to route and time of day, and which may also be void while streets are wet or icy. In practice, many localities do not enforce width restrictions for local moves. However, if an accident occurs to an over-width vehicle lacking a permit, the operator may find himself in a difficult position.

Drivers of full trailers should be sure of their route so as not to get into blind streets. The average driver cannot back such a rig any distance, or turn it around in restricted quarters. It may be necessary to disconnect the towing truck, turn that around, and tow the trailer out backwards. One or two men might be able to steer the trailer by manipulating the drawbar. If the load is a revolving shovel, it can pick up the tongue with a chain and steer by swinging. A jeep or wheel tractor following the trailer can sometimes manipulate the drawbar successfully.

Another method of getting out of a blind street is to unload the machine and drag the empty trailer sideward until it faces the exit.

Trailer routings must also be planned carefully to avoid underpasses too low to get through, and bridges too weak to cross over. Long low trailers can also get hung up crossing banked highways or railroad tracks and on sharp hilltops on country roads.

PUMPING. Dewatering of excavations commonly requires use of pumps.

If the water is small in volume or contains a heavy load of mud or other solids, a diaphragm pump is preferred. For larger quantities of water a centrifugal pump is needed. Both light hand-carry and heavier wheel models of centrifugals are available.

Most satisfactory results are obtained when the capacity of the pump or pumps is substantially greater than the inflow, particularly when there is a large volume of standing water.

Setting Up. The pump should be level and placed as near water level as possible, as centrifugals can push more strongly than they can pull. The table of output in Figure 78 shows the loss in volume which occurs as a pump is raised above the water. The capacity of a pump is greatest when total lift is low.

STANDARD TABLES

FOR SELF-PRIMING CENTRIFUGAL PUMPS

The following tables give capacities in gallons per minute

Model 4-M, 1½″

Total Head Including Friction	Height of Pump Above Water 10 Ft.	20 Ft.
15 ft.	67	
20 ft.	66	
25 ft.	65	47
30 ft.	63	47
40 ft.	54	44
50 ft.	37	34
55 ft.	25	25

Model 10-M, 2″

25 ft.	166	
30 ft.	164	115
40 ft.	157	113
50 ft.	145	107
60 ft.	122	97
70 ft.	85	75

Model 15-M, 3″

30 ft.	250	170
40 ft.	230	165
50 ft.	200	155
60 ft.	160	138
70 ft.	110	100

Model 20-M, 3″

Total Head Including Friction	Height of Pump Above Water 10 Ft.	20 Ft.
30 ft.	333	235
40 ft.	310	230
50 ft.	275	220
60 ft.	220	195
70 ft.	160	155
80 ft.	90	90

Model 30-M, 4″ light

30 ft.	500	350
40 ft.	495	345
50 ft.	475	340
60 ft.	450	325
70 ft.	415	300
80 ft.	355	270
90 ft.	250	215
100 ft.	100	100

Model 40-M, 4″ heavy

25 ft.	665	
30 ft.	660	475
40 ft.	645	465
50 ft.	620	455

Total Head Including Friction	Height of Pump Above Water 10 Ft.	20 Ft.
Model 40-M, 4″ heavy		
60 ft.	585	435
70 ft.	535	410
80 ft.	465	365
90 ft.	375	300
100 ft.	250	195
110 ft.	65	50
Model 90-M, 6″		
25 ft.	1500	
30 ft.	1480	1050
40 ft.	1430	1020
50 ft.	1350	970
60 ft.	1225	900
70 ft.	1050	775
80 ft.	800	600
90 ft.	450	365
100 ft.	100	100
Model 125-M, 8″		
25 ft.	2100	1570
30 ft.	2060	1560
40 ft.	1960	1520
50 ft.	1800	1450
60 ft.	1640	1360
70 ft.	1460	1250
80 ft.	1250	1110
90 ft.	1020	940
100 ft.	800	710
110 ft.	570	500

Condensed from "Contractors' Pump Standards.'

Fig. 78. Pump capacities

The pump should be supported on a platform, or on boards as in Figure 79. Otherwise vibration, and softening or washing of the ground, may cause it to settle off level, or even fall into the pit.

A strainer should always be used when there is a possibility of sucking up stones or other objects which might damage or clog the line; and it is a good practice to use one whenever possible.

Use of a foot valve is optional. If the hose diameter is small, the lift low, the pumping steady, and the inlet opening unlikely to be exposed, it is not needed. On high lifts, particularly where the pump is worn and the inlet valve defective, a foot valve is very desirable.

Foot valves are not particularly dependable, being more subject to jamming by trash or mud than parts of the pump proper. In addition, the end of the inlet hose is frequently horizontal or nearly so, and some valves do not function well in this position.

Inlet Protection. The inlet should be well below the surface of the water—about six times the inside diameter of the

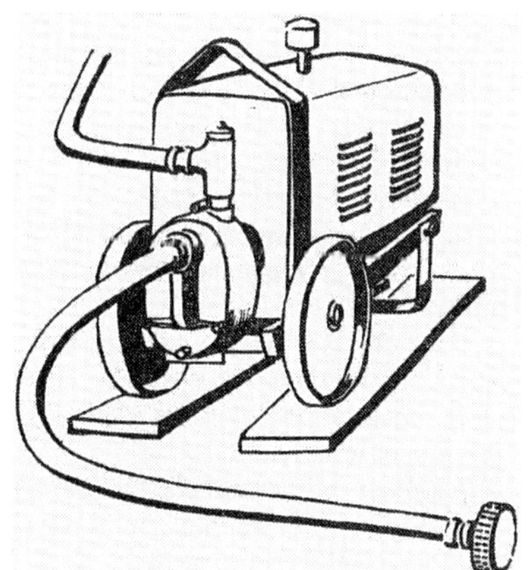

Fig. 79. Centrifugal water pump

hose when possible. This is frequently not practical, particularly when the place must be pumped dry.

At lesser depths a whirlpool may form over the inlet, and air enter through its center in sufficient quantity to form an air lock and cause the pump to lose its prime. Such a vortex will not form if the end of the hose is vertical, and is most likely to occur if it is horizontal.

The whirlpool may be prevented by digging a sump pit to lower the inlet; by floating a square or round piece of wood two or more feet in diameter over the inlet, or by bolting a roof over the strainer.

The inlet should not be allowed to rest in soft mud. It is liable to pick up abrasive material which will cause rapid wear of the pump; and may sink in sufficiently to reduce the intake, or choke it off entirely. It is particularly apt to be choked when the pump is not running, and mud can slump into the hollow it has dug by sucking in soil with the water.

This can be prevented by resting the strainer in a wood box, or on a wide board. A light metal plate may be fastened permanently to the under side of the strainer.

When leaves, grass, or other vegetable debris are in the water, a strainer may become plugged every few seconds. The best cure for this is to make an additional strainer of wire of sufficiently fine mesh to catch the bulk of the trash, with a large enough area so that the water will move through it at too low a velocity to hold the trash against it.

Even if too small for self-cleaning, such an additional screen will clog less often, and be more easily cleaned, than the standard strainer.

A riskier but less laborious method is to remove the hose strainer, and allow the trash to go through the pump.

Air in Inlet. A contractors' centrifugal pump is built to handle solid water, and pumps air with difficulty. When the pump and inlet hose are empty, the pump must be filled with water (primed) before starting, and will then work for a while slowly drawing the air out of the inlet pipe. When the water is sucked into the pump its efficiency rises abruptly. If air is permitted to enter the intake, the pump will lose its grip, and have to again slowly work up enough vacuum to lift the water into it. This process is fairly quick in new pumps, but in worn ones is often a slow process if the lift is at all high.

Very often failure of a pump to develop sufficient vacuum to prime itself is due to air leaks in the inlet line. These are most often past the gaskets at hose couplings, but may be at any spot on the inlet side. This includes the hoses themselves; gasket between the inlet elbow and the body, cracks in the inlet or body, threaded fittings, loose or worn shaft packings, or lack of grease in a shaft bushing.

Such leaks are hard to detect, as, if the pump is worn or the lift is high, a very small leak will do the damage. Wet clay should be smeared over all suspicious spots on the hose, couplings, gaskets, or casing parts. This makes a good temporary seal for small leaks; and will show up larger ones by disappearing into them. Leaks along the shaft will generally drip when the pump is stopped.

Gauges. The pump casing should be checked for cracks. Gaskets and connector threads should be smeared with pipe dope or Permatex #2. Installation of a vacuum gauge on the inlet side is an excellent idea, as it will immediately indicate the nature of the trouble. Low vacuum indicates leakage, a worn pump, or lack of prime water, and high vacuum a plugged inlet—usually the strainer, but sometimes a stuck valve. A fairly normal vacuum, with little or no production, indicates back pressure from too high an outlet or a plugged line.

A vacuum gauge may be calibrated in pounds or inches. If pounds, it measures the difference between air pressure inside and outside of the inlet chamber in pounds per square inch. If inches, it indicates the height of a column of mercury which could be supported by this difference in pressure.

Figure 80 contains a table for conversion of inches of vacuum into the number of feet that water should rise in the inlet

TABLE CONVERTING INCHES VACUUM INTO FEET SUCTION

To convert inches vacuum into feet, multiply by 1.13.

Inch Vac.	Feet	Inch Vac.	Feet	Inch Vac.	Feet
1	1.13	11	12.47	21	23.81
2	2.27	12	13.61	22	24.95
3	3.41	13	14.74	23	26.08
4	4.54	14	15.88	24	27.22
5	5.67	15	17.01	25	28.35
6	6.80	16	18.14	26	29.48
7	7.94	17	19.28	27	30.62
8	9.07	18	20.41	28	31.75
9	10.21	19	21.55	29	32.89
10	11.34	20	22.68		

CONVERSION FACTORS FOR PRESSURE AND VACUUM

Feet head (water) × .433 = pounds pressure
Pounds pressure (water) × 2.31 = feet head
Feet head (brine, sp. gr. = 1.2) × .52 = pounds pressure
Pounds pressure (brine) × 1.92 = feet head
Feet head (gasoline, sp. gr. = .75) × .325 = pounds pressure
Pounds pressure (gasoline) × 3.08 = feet head
Inches of mercury × 1.132 = feet head of water
Feet suction lift of water × .882 = inches of mercury

Courtesy of *Contractors' Pump Bureau.*

Fig. 80. Conversion tables

hose, and data for some other calculations.

A primed pump starting with an empty hose will increase its vacuum slowly, until the water rises into the pump and displaces the air. The increased efficiency of pumping water instead of air will cause the vacuum to jump to a higher figure than that needed to pull the water, after which it should drop to normal and remain steady as long as conditions are unchanged. As the water level drops as a result of pumping, the vacuum increases.

If the hose is full, vacuum will rise very rapidly to full pumping level.

Outlet or Discharge Line. If the outlet hose is not long enough to reach a disposal point, a level or down hill extension may be made by means of a wood flume, a ditch, or parallel dikes.

If the discharge hose end is under water, the pump is stopped, and the inlet and foot valves leak or are lacking, water will siphon from the higher level through the pump and out the inlet. On a high lift, the force may be sufficient to turn the pump and engine backward. Otherwise the flow is leakage past the vanes.

If the discharge point is high, thirty or more feet above the pump, it may be advisable to install a check valve just above the pump, to relieve the pump of destructive water hammer when shut down.

If the outlet hose is long or crooked, pump capacity may be increased by using oversize hose. This reduces the velocity and friction of the water. Sharp turns or abrupt changes in size should be avoided through use of taper fittings and 30° or 45° elbows.

It is frequently desirable to reduce the output of the pump, as when the water level has been lowered, and pumping need only remove an inflow which is less than the pump capacity. Gasoline power permits throttling the engine down. If this does not afford sufficient reduction, and it is not convenient to stop and start the pump frequently, a gate valve may be put on the outlet connection, and output regulated by opening or closing it. This does not add materially to the strain on the pump, whereas throttling the intake line does.

Freezing. Special precautions are necessary in operating a pump in freezing weather.

Whenever the pump is to be shut down, the body and hoses should be drained.

The drain-plug at the bottom of the case, and the filler plug, should be removed. If the end of the discharge hose is above the pump and underwater, it should be lifted out to avoid siphoning water down to the pump.

The discharge hose can be emptied of water by holding it so that there is a continuous slope to the ends. If this is not practical, water can be worked out of it by raising it at the high end, and walking toward the other end, holding it up in a loop high enough to force the water over other high points. If the line is metal, it should be opened at its low points, and blown out if air is available.

If the inlet hose is light, it can be disconnected, pulled out of the water, and drained. If it is heavy, and has no foot valve, it may drain itself if the pump check valve is not air tight. If it does not drain, an opening should be made in the metal at its upper end. A petcock may be provided, or a vacuum gauge removed, or a new hole cut and threaded. The air admitted in this way will allow the water inside the hose to drop to the surrounding water level.

This opening must be sealed when the pump is used again.

If freezing is apt to be severe enough to form a plug of ice at water level in the hose, the opening at the top should be made large enough to permit pouring sufficient alcohol or fuel oil into the hose to protect it. If the hose is deep in the water, one dose should give protection for a long time if the pump is not used. Should the opening in the hose be close to the surface, alcohol may have to be renewed every few days.

If a foot valve is required for the pumping it will be necessary to lift the lower end of the hose out of the water to open the valve to drain it unless a "tickler" is installed. Any homemade device which will unseat the flap in the foot valve by pulling on a line or a rod will be adequate.

If the pump body does not drain completely it may be necessary to tip it to get all the water out, or to pour in enough alcohol to prevent the residue from freezing.

The use of salt or calcium chloride as an anti-freeze is not recommended, because of probable damage to the pump.

The engine should be protected by antifreeze in the regular manner.

In freezing weather, it is advisable to start the engine before priming the pump, although it should not be allowed to run dry longer than a minute or two. If the temperature is below 0° F., it may be advisable to use hot water or a weak anti-freeze mixture for priming.

If the pump is frozen, the engine will not turn over. Filling the tank with hot water or with a strong anti-freeze solution should free it.

The outlet hose should be checked for ice before starting. Even if it has been drained, a small amount of water may collect in a low spot and glue the walls together. A hose blocked in any manner is liable to blow out under working pressure. Such ice can be readily broken by tapping the hose with a hammer or a block of wood.

Thawing an ice blocked inlet hose is a major operation, and methods will depend on the facilities on hand.

Under extreme conditions, a pump may freeze up while running. Provided the inlet is deep in the water, this may be prevented by wrapping hoses and pump body with sacking or blankets, or by applying heat (engine exhaust, bonfire or large blowtorch) to the case. Flame should not be used if there are gasoline leaks anywhere, and caution should be used if the engine is oily or dirty.

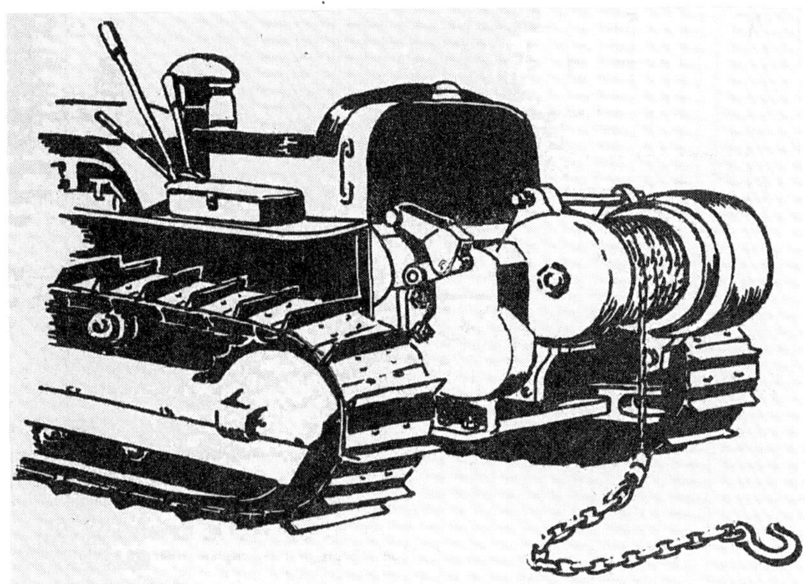

Fig. 81. Towing winch

WINCHES. The most powerful, most convenient, and best controlled line pull for uprooting stumps and trees, rescuing bogged machinery, and moving and setting heavy objects is obtained from winches. In excavation and clearing work these are usually mounted on crawler tractors, but may be on wheel tractors or trucks, or be portable units operated by a hand crank.

Crawler Mounted. A towing winch, Figure 80, consists of a heavy spool drum that is mounted on the back of a crawler tractor, and is driven by the power takeoff. It is controlled by the tractor engine clutch and the takeoff engagement lever. In addition, it may have a transmission giving rotation of the drum in either direction and, in large machines, permitting several speeds of rotation. A jaw clutch or neutral gear is used to disconnect the drum from the drive shaft, to allow it to turn freely when the cable is being removed. A brake is provided to slow or lock the drum when necessary.

The drum carries many layers of cable (wire rope), so that it has a greater spool diameter full than empty. It therefore reels in more slowly and powerfully on a bare drum than on a full one. On a bare drum, logging winches will give 50 to 100 percent more pull than the tractor itself; on a full drum, the same pull as the tractor or somewhat less.

Oil field winches, also mounted on crawler tractors, have tremendously stronger line pulls. Their great power is in part a side product of the very slow speeds required in delicate setting of heavy machinery, and can only be exerted when the tractor is firmly anchored and sufficiently heavy line is used.

A donkey or yarding winch has two drums with separate clutches and brakes. One drum is generally used for pulling in the load, and the other for pulling the cable out for the next load. They are extensively used for logging and hoisting, and can provide power for improvised cranes and drag scrapers.

On Wheel Tractors. A winch can be mounted on either end of a wheel tractor.

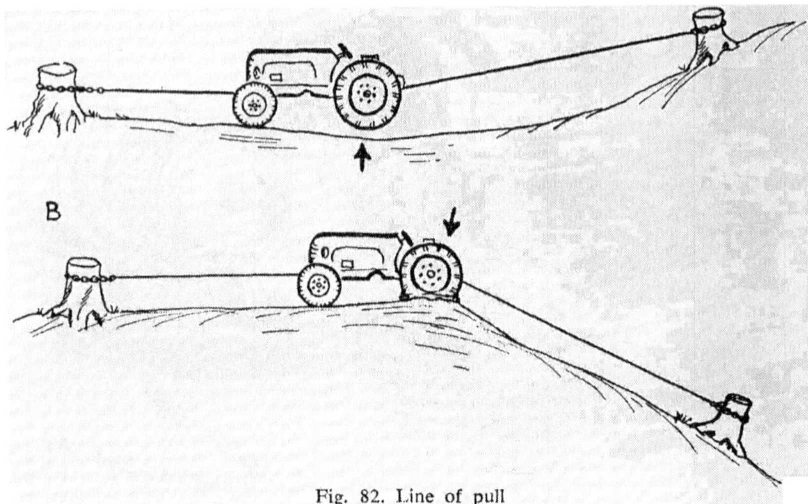

Fig. 82. Line of pull

It is usually necessary to anchor the machine for heavy pulls. The anchor chain or cable should be attached to the winch frame, or to a heavy member as near to it as possible, to reduce strain on the tractor.

An important consideration in the use of these winches is the fact that the cable will tend to take a straight line between the work and the anchor. In making a high pull, as in Figure 81 (A), the tightening cable may lift the tractor and turn it over sideways. In (B) the downward pull may blow the tires, unless the axle housing is blocked up.

If a wheel tractor is not anchored, a rear winch must be underwound, and care must be taken that the machine does not overturn through rising on the front, a danger which is particularly serious if the tractor is driven to move the load.

Winches that will pull loads up to fifty tons can be mounted on rather light wheel tractors.

On Trucks. Truck winches may be of the spool drum type, and may be mounted in the front bumper, on a flat bed body, or between the body and the cab. Gypsy spool winches or catheads are generally mounted vertically on the forward part of a flat body.

The principal handicap of a truck winch is the difficulty of maneuvering it into position for a straight pull. One or more pulleys may be required to obtain a proper direction of pull and a straight line onto the winch. The truck should have all the wheels blocked, or be anchored by a line from a frame member near the winch.

A gypsy, cathead, or capstan, Figure 82, does not carry cable. A hemp rope is looped around it two or three times, with one end attached to the work and the other end held by the operator. If the operator leaves it slack, the spool will turn inside the rope; if he pulls it tight, the working end of the rope will be pulled with great force. The slippage on the spool absorbs shocks that would break the rope and enables it to do quite heavy pulling, under exact control.

Hand Operated. Hand winches are turned by a hand crank, operating through one or more sets of reduction gears. Under most conditions it is not possible to make a full turn of the handle because it strikes obstructions, or passes through awkward positions. A large part of the work of winching consists in removing and replacing this handle, and if much work is to be done a ratchet handle should be purchased, or made up by adapting one from a heavy socket set.

The winch is usually equipped with a friction brake and a pawl that can be engaged to prevent it from turning backward when the handle is released.

Operation of these devices is tedious because of the number of crank turns which must be made to reel in the cable; and exhausting because of the force which must be applied to the handle to develop the rated pull of the winch. It is important that it be thoroughly lubricated.

Hand winches can be used in places inaccessible to power equipment, are comparatively inexpensive and are surprisingly powerful. Sometimes they can take out tougher stumps than a power winch of the same pull because the line can be left taut and tightened gradually or from time to time as the stump yields.

Their weight, with cable, may be from 75 to 300 pounds, so that carrying one of them any distance is at least a two man job. Carrying or dragging it with a truck or a small tractor is often advisable.

Hand winches are sometimes mounted on a truck, in which case they serve largely as a spool to carry cable, most of the pulling being done by the power of the truck. If the job is too heavy for the truck, it may be anchored or blocked and the work done with the winch handle.

For independent use the winch should have a V-shaped towbar, or a subframe by which it can be anchored. Blocks should be provided to build up a base in line with the pull as, if this is high, the winch will be lifted off the ground and will not be steady enough to allow turning the handle.

Cable (wire rope). A winch, whether power or hand operated, may hold one hundred, two hundred, or more feet of

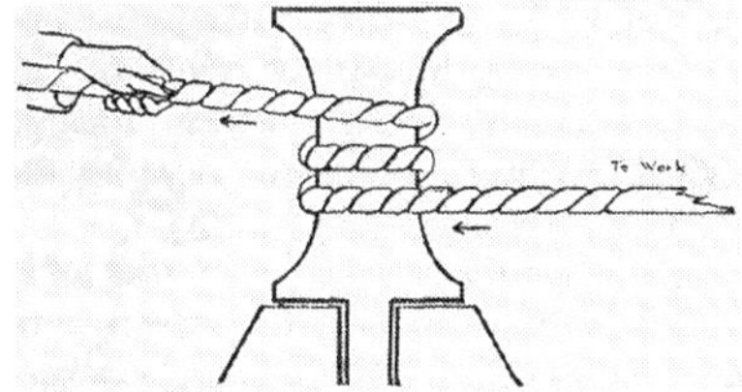

Fig. 83. Capstan or cathead

cable of a size proportionate to its power. Additional cable can be carried on a separate spool and connected to the winch cable by a choker device when needed.

In small sizes, the winch cable generally is fastened at the working end to a piece of alloy steel chain equipped with a round hook. Larger cables may be fastened directly, or through a swivel or single link, to a round hook, or a wide face cable grip hook. The cable is generally underwound on the drum, that is, leads from the work to the lower part of the drum. This gives best stability under heavy load.

If a double cable is so wrapped around the load that it cannot slide around it, great care must be taken to adjust it so that both ends share the strain equally, unless a single line is strong enough to take the entire pull alone.

A cable should never be bent sharply, pressed against a sharp edge, or dragged or chafed on rough surfaces.

Breakage. Cable or chain breakage is a serious danger to both operator and helpers. A cable particularly stretches under strain, and if it breaks suddenly, may whip with great force. The danger to the operator is greatest if the break is fairly near him. The cable used should be the best quality, in the largest size recommended by the winch manufacturer if the tractor is to be anchored for heavy pulls, and it should be inspected frequently for weak spots. If breakage is frequent, a guard screen should be placed behind the tractor seat.

Another danger inherent in the use of cable is cutting and tearing of the hands and clothing on broken wires. Preformed cable gives minimum trouble of this kind and should be used when possible. Leather palmed gloves are good protection for the hands.

Either hemp center or wire center cable may be used, according to preference or manufacturer's recommendation. Wire center is about 10 percent stronger, size for size, is stiffer, and is not as easily deformed by crushing. It is more difficult to handle, and when kinked or crushed is much harder to straighten. Standard 6 x 19 constructions are usually recommended.

It is good practice to work a winch at less than its maximum capacity, and to avoid anchoring the tractor unless absolutely necessary. Moderate loads give long life to cables and winch parts, and avoid severe catching on the drum. If the work is heavy, strain can be reduced by the use of pulleys and multiple lines.

Broken cables can be repaired by splicing, but the length of cable used in the splice, and the labor involved, may be too great to justify this method for the short cables ordinarily used in clearing land and moving machinery.

A rough repair may be made by trimming back the broken ends, overlapping them, fastening them with two or three cable clamps, for sizes up to half inch or five eighths, or with three or more for larger sizes. Or two interlocked loops may be made with clamps or any type of loop fastening.

Cables patched in this manner are weakened but may last a long time. The patch will not go through pulleys and is inconvenient in other ways.

Chain. A separate chain is often used on the load to save the winch line damage from short bends and kinking. A standard logging chain is composed of short straight links, carries a round hook on one end and a grabhook on the other. The round hook may be fastened to the chain by a ring, or a ring may be used instead of this hook.

Either the round hook or the ring can be used in chokers which are loops which tighten automatically when pulled. The hook is easier to attach and to detach, but may fall away from the chain when it is slack. The ring may be used by passing the grabhook through it and pulling from the grabhook end; or for stumping, the chain near the ring may be pulled through it to form a loop that is dropped over the stump.

The grabhook fits over any individual chain link, and will not slide along the chain. It is used to adjust the length of chain by increasing or decreasing the amount of double line, by moving it toward or away from the choker end.

Alloy steel chain weighs only about a third as much as standard chain in proportion to strength. If a crew is careful enough not to lose chains, and is conscientious enough not to abuse them by kinking or gross overloading, alloy chains will amply repay their much higher cost in reduced labor and fatigue, and by greater efficiency. It is recommended that they be dipped in bright red paint, so that they can be easily recognized and recovered if mislaid.

A broken chain is best repaired by having new links forged into it by a blacksmith. However, good field repairs may be made with a variety of patent repair links, or by shackles with removable pins. Links that have been stretched thin may not admit the proper size repair piece, and may have to be opened by supporting on a block with a hole in it, and spreading with a punch and hammer.

A broken chain may be used temporarily by making a square knot, and tying the end links together with wire or string. If there is not enough chain, a half knot should be used and the ends fastened with a bolt. A bolt may be used without any knot, but will pull apart more easily than a link.

Two grabhooks connected by a ring can be used for temporary repair and for shortening a chain.

Rigging stumps. Often, the principal

job of a winch is stump pulling. The stump line is generally a choker type which tightens its grip as the pull increases. In smaller sizes, chain is preferred because it is easier to carry, safer to handle, and more resistant to abuse. However, it is much heavier than cable for the same strength, and in large sizes is too weighty to be practical.

Figure 83 shows three ways of fastening a winch line to a stump. It is a good idea to cut a groove in the back of the stump, which will hold the line against squeezing off during the pull, and will delay its slipping off as the stump leans.

(A) is the easiest and most usual method, with a sharp kink in the chain and pull at the top center of the stump. (B) is a little harder to arrange, but adds a twisting effect to the direct pull, and since it does not bend the line as sharply it can be used with cable chokers. The overtop method, (C) gives the greatest leverage, but is the most likely to slide off.

Care should be exercised not to put loads on a line that is twisted or kinked, as it will be damaged. Cable is particularly sensitive. A chain can be easily checked for straightness as the links which are in one plane should lie in a straight line.

Pulling. The tractor should be placed facing directly away from the stump, with both brakes locked on. The winch jaw clutch should be released and the brake set to drag very slightly, and the cable pulled to the stump by hand. If the brake is not used, the drum may continue to spin after being pulled, and unwind and snarl the cable. If the winch will not freewheel, or the cable is very heavy, it is turned backwards by the engine, and it is convenient to have two men, one to operate the winch and the other to pull the cable. If no helper is available, the operator can stand near the winch while it turns, stripping the cable and coiling it on the ground until he thinks he has enough. He then stops the winch and drags the cable to the stump. The cable must then be whipped up and down and twists worked out to avoid kinking.

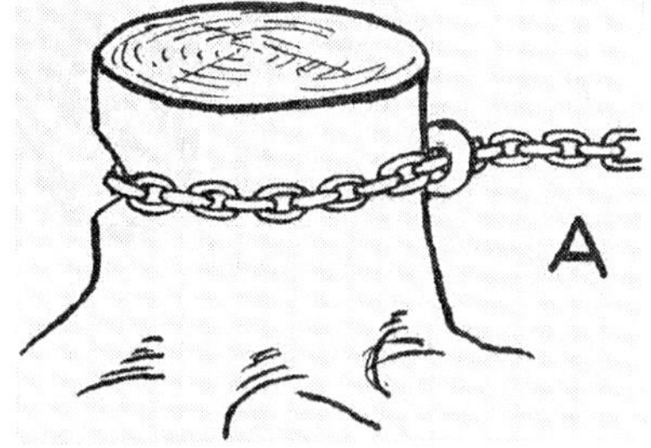

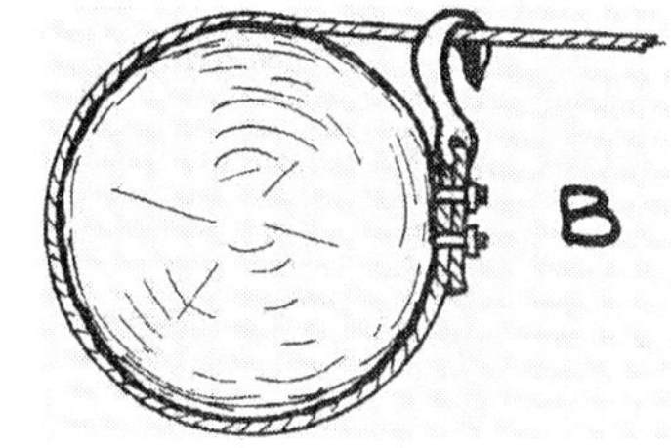

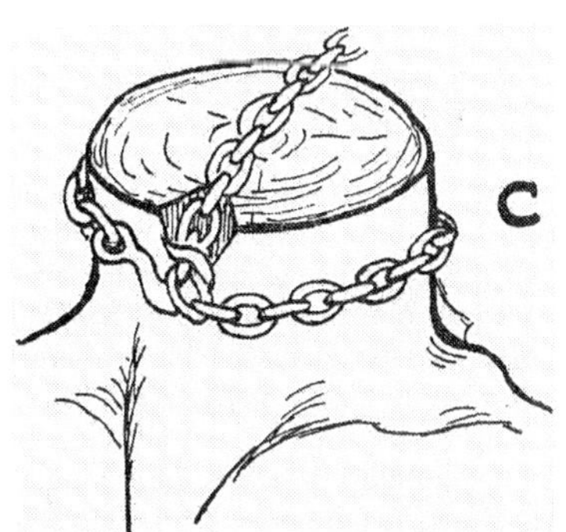

Fig. 84. Stump chokers

The winch cable may be put around the stump directly, hooked to a choker chain or cable, or may be run through one or more snatch blocks to step up its pull.

Power is applied to the winch and the cable is reeled in, care being taken to see that it feeds onto the drum properly. The

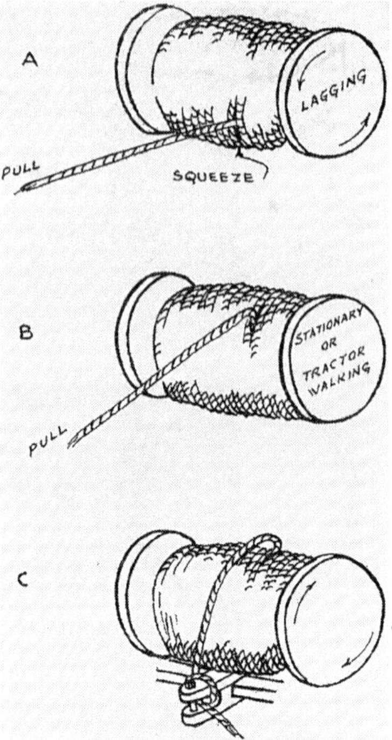

Fig. 85. Jammed cable

stump may come out or the tractor may be dragged backward. If the latter, the tractor may be anchored by a chain from the blade or front pull hook to a tree. Resistance to pull may also be increased by backing it against a log or bank, or by trying to pull the stump by tractor pull and allowing the tracks to spin until they have built mounds behind them. If the anchoring or blocking is effective, the stump will come out if nothing breaks or slips and the engine does not stall.

Jammed Cable. Using a nearly bare drum not only gives the greatest pull but reduces damage to the cable. If a long cable is wound smoothly onto a drum under moderate tension, and a heavy pull applied when it has built up several layers, the last wrap may squeeze between the wraps below, as in Figure 84 (A). This scrapes and wears the cable and jams it so that it will not spool off again. The best way to free it is to turn the drum until the catch is in the position shown in (B), and jerk it by hand or anchor the end and drive the tractor away. Or, in the same position on the drum, the cable may be given a couple of wraps around the drawbar, and the winch turned backward as in (C).

If the cable is wound unevenly onto the drum, with the wraps crossing each other at random, it cannot cut down between lower layers readily, but may put severe kinks in sections of cable that cross under it, and this cross wrapping may not entirely prevent it from squeezing in and sticking.

In spite of these difficulties, a long cable is desirable for general work. If reasonable effort is made to spool it in evenly while working, it will usually be rough enough to prevent excessive sticking, without too much bending or crushing.

Two-part Line. Where the distance to the load is less than half the cable length, a two-part line may be used by attaching a pulley (snatch block) to the stump and by running the line from the winch around the pulley and back to the drawbar. The useful strength of the cable, and the pull between the tractor and the stump are doubled.

The tractor may have to be backed against a heavy log or an outside anchor used in the manner to be described below. The tractor should not be anchored by the front pull hook while using a double line anchored on the drawbar, unless the manufacturer will state that it is strong enough to take the strain.

Resistance. A stump's resistance varies in different directions. If on a slope, downhill pull is most effective. Otherwise it should be pulled toward its strongest roots, as these are easier to bend than to pull apart, and can be dealt with more easily when the rest of the stump is loosened.

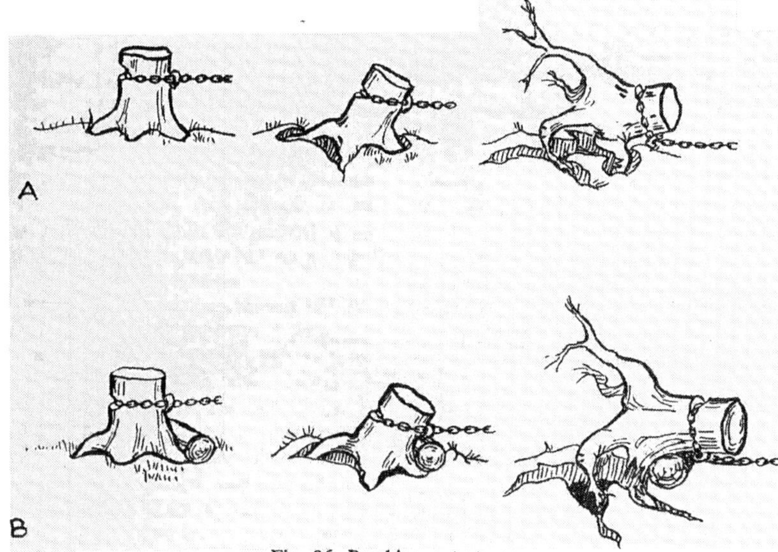

Fig. 86. Breaking out stump

The most obvious variable in stump resistance is its height. Greater height means greater leverage and easier pulling. Limiting factors are difficulty of high cutting, of fastening heavy chains at a height, and of the trunk breaking under pull.

A stump which yields to pull but will not break loose, can often be uprooted by moving it as far as possible, slacking off to allow it to settle back, and pulling again, repeating this process a number of times. This is most effective if done slowly and smoothly, whether with winch or traction.

This method is very effective with trees, as the trunk will bend with a whipping motion that exaggerates the force of both the pull and the snap back.

Chopping the roots on the side opposite the pull, while they are under maximum tension, weakens the resistance. A moderate amount of digging will generally expose the main lateral roots.

When a stump has been split by blasting, the pieces are most easily pulled away from the center, rather than across it.

Taproots. The presence of a taproot increases resistance of the stump. If the ground is hard, this root may be broken or pulled apart. If the ground is soft, or the wood very tough or pliable, the pivot point may crush and the root bend so that the pulling power is exerted directly against the length of the root, without benefit of leverage. In such a case, the upper roots of the stump may be torn up sufficiently so that an ax, or a special long chisel, can reach and cut the taproot. The cut should be made while pulling as tension makes the wood part more easily.

A buried stump is the hardest of all to pull and usually must be dug out.

On filled land, two separate systems of lateral roots may be found, one under the old ground level and the other near the surface, in which case it may be necessary to cut the trunk below the upper roots, in the same way as a taproot.

Pulling Clear. If the force is sufficient to uproot a stump, the roots opposite the pull break first, then those at the side, permit-

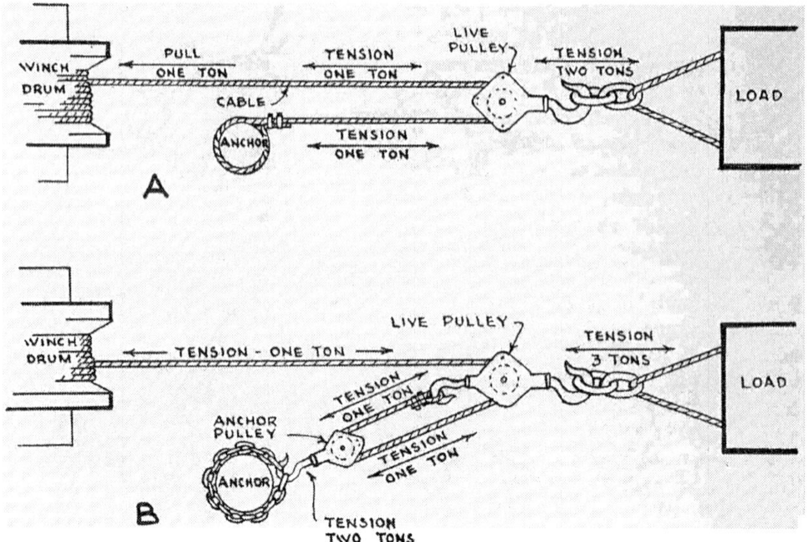

Fig. 87. Multiple lines

ting the stump to be pulled onto its side, as in Figure 86 (A). If the line does not slip off, the stump may be rolled and dragged out of the ground, but this often takes much more power than overturning the stump, and may be beyond the capacity of the machine.

If the stump will not come all the way, the line may be slacked, and a log placed against the stump, as in (B). This log will provide a new fulcrum and aid the breaking out, but it may have to be chained or staked in place.

Or the line may be taken off and the tractor moved to pull in the opposite direction, which should free it without difficulty. If a number of stumps are being pulled, all of them may be overturned one way, before pulling the tough ones in the opposite direction.

Half-uprooted stumps are easily knocked out by dozers, and may be left for them to save the trouble of re-rigging.

Multiple Lines. If pulling a load takes the full power of the tractor or winch, it may be advisable to use snatch blocks to rig multiple lines to obtain greater power at slower speed. These devices, also known as blocks and as pulleys, are pulleys set in frames that are provided with one or two round hooks or rings on swivel connections. For most field work, single pulley wheels, with a latch arrangement permitting insertion of a cable at the side, are best, as cables usually carry attachments too large for threading through rigid cases, which is a tedious job even when possible.

These blocks can be obtained in sizes to match any cable or stress. In large sizes, they are very heavy, and several men, a small wheel tractor, or light donkey winches, may be used to move them.

Typical riggings using pulleys are shown in Figure 87. If the lines are approximately parallel, and the pulley bushings lubricated, each additional line will add about 90 per cent to the single line pull. This puts no extra strain on the winch cable, but the chokers holding the blocks must take the combined pull of all the lines fastened to them.

The advantage obtained from the use of

a block is decreased when the lines are not parallel, becoming zero when the angle between the lines is 114°. Still wider angles result in loss of power.

The number of lines that can be put on one stump may be determined by the number of blocks on hand, the space available for fastening them, the strength of the available anchorage, or the amount of cable. Light machines may use six or eight blocks on a heavy stump.

Rigging is simplified by the use of series blocks. The tractor line passes around a block pulley to an anchor, doubling the pull at the block frame. This frame is attached to a heavier line which passes around another pulley to an anchor. The second pulley is fastened to the stump choker, and is pulled with almost four times the power of the winch.

Rigging. Multiple blocks require care in rigging and pulling. Two anchors are better than one as they spread the cables over a wider space, where they are less apt to interfere with each other. Each block is best fastened to a separate choker, but one may be fastened to each end of a chain passed behind the stump if it is strong enough to take a double pull; or to the ends of a chain given one turn around the stump. It is good practice to notch the stump for each chain used so that the chains and blocks will not slide into each other as it yields.

Rigging is done with the lines slack. When they are pulled tight an inspection should be made to make sure that no pulley latches have fallen open, as a pull on an open block will bend it and cut the cable; that no pulleys are jammed with debris, or liable to pull into each other, and that no chain hooks have become disengaged. As the line is wound in, all blocks should be watched to make certain that they do not collide.

Anchors. It is often a question of whether the stump or the anchor will yield. Anchor lines should be as low as possible, and stump lines high. It may be best to pull the largest stumps first, using several smaller ones for anchorage if necessary. In a clean clearing job, there is always one last stump for which there is no anchor, and if it is small, it may be pulled out directly; or in any case it will respond to less elaborate artificial anchors than a large one. On the other hand, a large stump will be a dependable anchor, and will prevent the need of frequent re-rigging when anchors pull out of the ground.

The final stump may be pulled by use of a living tree as an anchor. A choker should not be used under any circumstances on a tree which is to be preserved, and padding and blocks should be used to protect it.

If no anchor is available one may be made, ground conditions permitting, by digging a trench at right angles to the direction of pull, with a steep side pierced by an inclined slot toward the pull. A log is placed in it, and the cable is anchored to the center of the log and led up through the slot toward the work.

Advantages of Blocks. Load or stump pulling with a winch and blocks takes more time and care than direct winch pull, but results are generally more satisfactory. Jerks and jars which are destructive to machinery and cables are largely eliminated. Lighter cable may be used and a sufficient number of lines will reduce the tension on any one so much that squeezing and crushing on the drum will not occur.

A block can also be anchored to the side of a work area, and used to guide a cable between the winch and the objects being pulled. This makes it unnecessary to change the setting of the winch for each change in direction.

GLOSSARY

A-frame. An open structure tapering from a wide base to a narrow load-bearing top.

Apron. The front gate of a scraper body.

A short ramp with a slight pitch.

Back Haul. A line which pulls a drag scraper bucket backward from the dump point to the digging.

Bank. Specifically, a mass of soil rising above a digging or trucking level. Generally, any soil which is to be dug from its natural position.

Bank Gravel. A natural mixture of cobbles, gravel, sand, and fines.

Bank Measure. Volume of soil or rock in its original place in the ground.

Bank Yards. Yards of soil or rock measured in its original position, before digging.

Bell Crank. A lever whose two arms form an angle at the fulcrum, or a triangular plate hinged at one corner.

Belt Conveyor. An endless pulley-driven belt supported on rollers, which transports material placed on its upper surface.

Berm. An artificial ridge of earth.

Bit. The part of a drill which cuts the rock.

Blade. Usually a part of an excavator which digs and pushes dirt but does not carry it.

Block. A pulley and its case.

Boom. In a revolving shovel, a heavy beam hinged to the deck front, supported by cables. Any heavy beam which is hinged at one end and carries a weight-lifting device at the other.

Boom, live. A shovel boom which can be lifted and lowered without interrupting the digging cycle.

Borrow Pit. An excavation from which material is taken to a nearby job.

Bowl. The bucket or body of a carrying scraper. Sometimes the moldboard or blade of a dozer.

Bucket Sheave (Padlock sheave). A pulley attached to a shovel bucket, through which the hoist or drag cable is reeved.

Bull wheel. A large driving wheel or sprocket.

Bulldozer. A tractor equipped with a front pusher blade.

Cable. Rope made of steel wire.

Carbide Bit. A steel bit which contains inserts of tungsten carbide.

Cave. Collapse of an unstable bank.

Centrifugal Force. Outward force exerted by a body moving in a curved line. It is the force which tends to tip a car over in going around a curve.

Cherry Picker. A small derrick made up of a sheave on an A-frame, a winch and winch line, and a hook. Usually mounted on a truck.

Chock. A block used under and against an object to prevent it from rolling or sliding.

Choker. A chain or cable so fastened that it tightens on its load as it is pulled.

Chuck. The part of a drill which rotates the steel.

Circle. In a grader, the rotary table which supports the blade and regulates its angle.

Clamshell. A shovel bucket with two jaws which clamp together by their own weight when it is lifted by the closing line.

Closing Line (Digging line). The cable which closes the jaws of a clamshell bucket.

Clutch. A device which connects and disconnects two shafts which revolve in line with each other.

Clutch Brake. A device to slow the jackshaft when a clutch is released, to permit more rapid gear shifting.

Clutch, engine (Flywheel clutch). A friction clutch in an engine flywheel.

Collar. A sliding ring mounted on a shaft so that it does not revolve with it. Used chiefly in clutches and transmissions.

Collaring. Starting a drill hole. When the hole is deep enough to hold the bit from slipping out of it, it is said to be collared.

Compaction. Reduction in bulk of fill by rolling, tamping, or soaking.

Conveyor. A device which transports material by belts, cables, or chains.

Countershaft. A shaft which receives power from a parallel mainshaft, and transmits it to another part of the mainshaft or to working parts.

Counterweight. A "dead" or nonworking load attached to one end or side of a machine to balance weight carried on the opposite end.

Crawler. One of a pair of roller chain tracks used to support and propel a machine, or any machine mounted on such tracks.

Crowd. The process of forcing a bucket into the digging, or the mechanism which does the forcing. Used chiefly in reference to machines which dig by pushing away from themselves.

Crown. The elevation of a road center above its sides.

Cutter. On a hydraulic dredge, a set of revolving blades at the end of the suction line.

Cutting. Excavating.

Lowering a grade.

Cycle, digging. Complete set of operations a machine performs before repeating them.

Cylinder. In hydraulic systems, a hollow cylinder of metal, containing a piston, piston rod and end seals, and fitted with a port or ports to allow entrance and exit of fluid.

Dewatering. Removing water by pumping, drainage, or evaporation.

Diaphragm. A flexible partition between two chambers.

Digging Line. On a shovel, the cable which forces the bucket into the soil. Called crowd in a dipper shovel, drag in a pull shovel and dragline and closing line in a clamshell.

Dip. The slope of layers of soil or rock.

Dipper Stick. A name for the standard revolving shovel (dipper shovel), and for the straight shaft which connects the bucket with the boom.

Dipper Trip. A device which unlatches the door of a shovel bucket to dump the load.

Diversion Valve. A valve which permits flow to be directed into any one of two or more pipes.

Dog. A heavy duty latch.

Dolly. A unit consisting of a draw tongue, an axle with wheels, and a turntable platform to support a trailer gooseneck.

Donkey. A winch with two drums which are controlled separately by clutches and brakes.

Dozer. Abbreviation for bulldozer or shovel dozer.

Draft. Resistance to movement of a towed load.

Drag. Pulling a bucket into the digging, or the mechanism by which the pulling is done or controlled.

Drag Brake. On a revolving shovel, the brake which stops and holds the drag (digging) drum.

Dragline. A revolving shovel which carries a bucket attached only by cables, and digs by pulling the bucket toward itself.

Drag Scraper. A digging device consisting of a bottomless bucket working between a mast and an anchor.

Drum. A rotating cylinder with side flanges used for winding in and releasing cable.

Fairlead. A device which lines up cable so that it will wind smoothly onto a drum.

Feather. To blend the edge of new material smoothly into the old surface.

Feed. A mechanism which pushes a drill into its work.

Feed Travel. The distance a drilling machine moves the steel shank in traveling from top to bottom of its feeding range.

Fifth Wheel. The weight-bearing swivel connection between highway-type tractors and semitrailers.

Fines. Clay or silt particles in soil.

Float. In reference to a dozer blade—to rest by its own weight, or to be held from digging by upward pressure of a load of dirt against its moldboard.

Fluid Clutch. A hydraulic coupling which does not increase torque.

Flume. An artificial channel, often elevated above the ground, used for carrying fast flowing water.

Foot Pins. The hinge which attaches the boom to a revolving shovel.

Foot Valve. A check valve in the inlet end of a pump suction hose.

Four-Part Line. A single rope or cable reeved around pulleys so that four strands connect the fixed and the movable units.

Frost. Frozen soil.

Fulcrum. A pivot for a lever.

Full Trailer. A towed vehicle whose weight rests entirely on its own wheels or crawlers.

Gooseneck. An arched connection, usually between a tractor and a trailer.

Gravel. Rock fragments from 2 mm to 64 mm in diameter. Or a mixture of such gravel with sand, cobbles, boulders, and not over 25 percent of fines.

Grizzly. A coarse screen used to remove oversize pieces from earth or blasted rock.

Grouser. A ridge or cleat across a track shoe, which improves its grip on the ground.

Head. Height of water above a specified point. The back-pressure against a pump from a high outlet.

Hog Box. A concrete box in which water and dirt are mixed to be pumped to a fill.

Holding Line. The hoist cable for a clamshell bucket.

Horse. A saw horse or other simple frame or support.

Jack Boom. A boom which supports sheaves between the hoist drum and the main boom in a pull shovel or a dredge.

Jackshaft. A short drive shaft, usually connecting a clutch and transmission.

Jackknife. A tractor and trailer assuming such an angle to each other that the tractor cannot move forward.

Jaw. In a clutch, one of a pair of toothed rings, the teeth of which face each other.

Jib Boom. An extension piece hinged to the upper end of a crane boom.

Knife. Cutting edge of a digging machine.

Monitor. In hydraulicking, a high pressure nozzle mounted in a swivel on a skid frame.

Moldboard. A curved surface of a plow, dozer or grader blade, or other dirt mover, which gives dirt moving over it a rotary, spiral, or twisting movement.

Muck. Mud rich in humus.

Finely blasted rock, particularly from tunnels.

Mud. Generally any soil containing enough water to make it soft.

In rotary drilling, a mixture of water with fine drill cuttings and added material, which is pumped through the drill string to clean the hole and cool the bit.

Mudcapping. Blasting boulders or other rock by means of explosive laid on the surface and covered with mud.

One Part Line. A single strand of rope or cable.

Open-Cut. A method of excavation in which the working area is kept open to the sky. Used to distinguish from cut-and-cover and underground work.

Oscillation. Independent movement through a limited range, usually on a hinge.

Outrigger. An outward extension of a frame which is supported by a jack or block. Used to increase stability.

Overburden. Soil or rock lying on top of a pay formation.

Overhead Shovel. A tractor which digs at one end, swings the bucket overhead, and dumps at the other end.

Overwinding. A rope or cable wound and attached so that it stretches from the top of a drum to the load.

Pan. A carrying scraper.

Part Swing Shovel. A shovel in which the upper works can rotate through only part of a circle.

Parts of Line. Separate strands of the same rope or cable used to connect two sets of sheaves.

Pass. A working trip or passage of an excavating or grading machine.

Paving Breaker. An air hammer which does not rotate its steel.

Peg Point (Steady point). A pointed bar in a slide clamp. Used to brace a machine during work.

Pit. Any mine, quarry, or excavation area worked by the open-cut method to obtain material of value.

Pitch. The slope of a surface or tooth relative to its direction of movement.

Pitch Arms (Pitch braces) (Pitch rods). Rods, usually adjustable, which determine the digging angle of a blade or bucket.

Platforms (Pontoons) (Mats). Movable wood platforms used in sets to support machines on soft ground.

Plumb Bob. A pointed weight hung from a string. Used for vertical alignment.

Pond. In dredge work, an area where water is held long enough to allow fine soil particles to settle.

Port. Left side of a boat.

Power Arm. The part of a lever between the fulcrum and the point where force is applied.

Power Control Unit. One or more winches mounted on a tractor and used to manipulate parts of bulldozers, scrapers, or other machines.

Power Takeoff. A place in a transmission or engine to which a shaft can be so attached as to drive an outside mechanism.

Pre-Selective. An arrangement by which a gear lever can be moved, but the resulting speed shift will not take place until the clutch or the throttle is manipulated.

Primary Excavation. Digging in undisturbed soil, as distinguished from rehandling stockpiles.

Prime. To provide means to start a process, as to supply sufficient water to a pump to enable it to start pumping.

In blasting, to place a detonator in a cartridge or charge of explosive.

Prime Mover. A tractor or other vehicle used to pull other machines.

Puff Blowing. Blowing chips out of a hole by means of exhaust air from the drill.

Pumping. In scraper operation, raising and lowering the bowl rapidly to force a larger load into it.

Quarrying. Removal of rock which has value because of its physical characteristics.

Ram. A hydraulic cylinder.

Ram, One Way or Single Acting. A hydraulic cylinder in which fluid can be supplied to only one end so that the piston can be moved only one way by power.

Ram, Two Way or Double Acting. A hydraulic cylinder in which fluid can be supplied to either end, so the piston can be moved by power in two directions.

Ramp. An incline connecting two levels.

Reeving. Threading or placement of a working line.

Retract. The mechanism by which a dipper shovel bucket is pulled back out of the digging.

Revolving Shovel. A digging machine in which the upper works can revolve independently of its supporting unit.

Reversing Clutch. A forward-and-reverse transmission which is shifted by a pair of friction clutches.

Rifling. Forming a spiral thread on the wall of a drill hole, which makes it difficult to pull out the bit.

Rig. A general term, denoting any machine. More specifically, the front or attachment of a revolving shovel.

Ripper. A rooter.

Rock. The hard, firm, and stable parts of earth's crust.

Any material which requires blasting before it can be dug by available equipment.

Rooter. A towed machine equipped with teeth, used primarily for loosening hard soil and soft rock.

R.P.M. or r.p.m. Revolutions per minute.

Saddle Block. In a dipper shovel, the boom swivel block through which the stick slides when crowded or retracted.

Safety Clutch. A clutch which will slip under loads which might damage the machine.

Sand. A loose soil composed of particles between 1/16 mm and 2 mm in diameter.

Rock chips and other waste produced by drilling action.

Scarifier. An accessory on a grader, roller, or other machine, used chiefly for shallow loosening of road surfaces.

Scraper (Pan). A machine which has a cutting edge, a bowl, and an apron, and which digs, transports, and spreads soil.

Selective Digging. Separating two or more types of soil while digging them.

Semi-Trailer. A towed vehicle whose front rests on the towing unit.

Shank. A bar or standard which connects a rooter tooth with the frame.

Sheave. A grooved wheel used to support cable or change its direction of travel. Pronounced "shiv."

Sheave Block. A pulley, and a case provided with means to anchor it.

Shipper Shaft. In a dipper shovel, the hinge on which the stick pivots when the bucket is hoisted.

Shoe. A ground plate forming a link of a track, or bolted to a track link.

A support for a bulldozer blade or other digging edge to prevent cutting down.

A cleanup device following the buckets of a ditching machine.

Shoulder. The part of a road between the pavement and the gutter.

Shuttle. A back and forth motion of a machine which continues to face in one direction.

Sidecasting. Piling spoil alongside the excavation from which it is taken.

Side Hill. A slope which crosses the line of work.

Silt. A soil composed of particles between 1/256 mm and 1/16 mm in diameter.

Siphon. A tube or pipe through which water flows over a high point by gravity.

Slackline Cableway. A cable excavator having a track cable which is loosened to lower the bucket, and tightened to raise it.

Sling. A lifting hold consisting of two or more strands of chain or cable.

Sluice. A steep, narrow waterway.

Snag Boat. A boat equipped with a hoist and grapple for clearing obstacles from the path of a dredge.

Snatch Block. A pulley in a case which can be easily fastened to lines or objects by means of a hook, ring, or shackle.

Soil. The loose surface material of the earth's crust.

Spiraling. Rifling.

A drill hole twisting into a spiral around its intended center line.

Spoil. Dirt or rock which has been removed from its original location.

Spool. To wind in a winch cable.

Spot Log. A log or marker placed to show a truck driver the spot where he should stop to be loaded.

Spotting. Placing trucks for loading or dumping.

Spuds. On a dredge, steel tubes pointed at the bottom and provided with lifting tackle at the top which are used to hold and to move the dredge.

Starboard. Right side of a boat.

Starter. In drilling, a short steel used to start a drill hole.

Steady Point (Peg point). A pointed steel bar which can be locked in a clamp, and is used to brace a drill frame against the ground.

Steel. In air hammers, the hollow or solid steel bar which connects the hammer with the cutting tool.

Steel Centralizer. On a wagon drill, a guide to hold the starting steel in proper alignment.

Steel Changes. The difference in length between successive steels used in drilling one hole.

Steel Puller. A hinged clamp on the bottom of a hand drill.

Steering Brake. A brake which slows or stops one side of a tractor.

Steering Clutch. A clutch which can disconnect power from one side of a tractor.

Stick or Handle. In a dipper shovel or pull shovel, a rigid bar hinged to the boom and fastened to the bucket.

Stockpile. Material dug and piled for future use.

Strip. Remove overburden or thin layers of pay material.

Stripping. Removal of a surface layer or deposit.

Stripping Shovel. A shovel with an especially long boom and stick which enables it to reach further and pile higher.

Suck. The shape of the bottom of a cutting edge or tooth which tends to pull it into the ground as it is moved.

Suction. Resistance of objects to being lifted out of mud. Caused by atmospheric pressure.

Swing. In revolving shovels, to rotate the shovel on its base.

In churn drills, to operate a string of tools.

Swing Angle. The distance in degrees which a shovel must swing between digging and dumping points.

Tail Anchor. The anchor for a track cable, or the turn point for a backhaul line in a cable excavator.

Tail Swing. The clearance required by the rear of a revolving shovel.

Tailgate. The hinged rear wall of a dump truck body.

The hinged or sliding rear wall of a scraper bowl.

Tandem. A pair in which one part follows the other.

Taproot. A big root which grows downward from the base of a tree.

Three Part Line. A single strand of rope or cable doubled back around two sheaves so that three parts of it pull a load together.

Tight. Soil or rock formations lacking veins of weakness.

Blasts or blast holes around which rock cannot break away freely.

Tilting Dozer. A bulldozer whose blade can be pivoted on a horizontal center pin to cut low on either side.

Toe. The projection of the bottom of a face beyond the top.

Tongue. Drawbar of a towed vehicle.

Topsoil. The topmost layer of soil. Usually refers to soil containing humus and which is capable of supporting a good plant growth.

Torque. The twisting force exerted by or on a shaft, without reference to the speed of the shaft.

Torque Converter. A hydraulic coupling which utilizes slippage to multiply torque.

Track. One of a pair of roller chains used to support and propel a machine. It has an upper surface which provides a track to carry the wheels of the machine, and a lower surface providing continuous ground contact.

Traction. The total amount of driving push of a vehicle on a given surface.

Tractor. A motor vehicle on tracks or wheels used for towing or operating vehicles or equipment.

Tractor Loader. A tractor carrying a bucket which can dig, and elevate to dump at truck height.

Trailer (Full trailer). A towed carrier which rests on its own wheels both front and rear.

Truck Roller. In a crawler machine, the small wheels which are under the track frame and which rest on the track.

Two Part Line. A single strand of rope or cable doubled back around a sheave so that two parts of it pull a load together.

Underwinding. A rope or cable wound and attached so that it stretches from the bottom of a drum to the load.

Wagon. A full trailer with a dump body.

Walking Dragline. A dragline shovel which drags itself along the ground by means of side mounted shoes.

Water Table. The surface of underground, gravity-controlled water.

Winch. A drum, and means for rotating and controlling it by power or hand.

Window Pipe. A dredge discharge pipe with one or more openings in the bottom.

Windrow. A ridge of loose dirt.

Work Arm. The part of a lever between the fulcrum and the working end.

Working Cycle. A complete set of operations. In an excavator, it usually includes loading, moving, dumping, and returning to the loading point.

Worm. A gear with spiral threads cut in a cylinder.

CPSIA information can be obtained
at www.ICGtesting.com
Printed in the USA
BVHW031006010522
635837BV00010B/249